Enteroviruses

Omics, Molecular Biology, and Control

https://doi.org/10.21775/9781910190739

Edited by

William T. Jackson

Department of Microbiology and Immunology, University of Maryland School of Medicine, Baltimore, MD, USA

and

Carolyn B. Coyne

Department of Pediatrics, University of Pittsburgh, Pittsburgh, PA, USA

Caister Academic Press

Caister Academic Press
Norfolk, UK

www.caister.com

British Library Cataloguing-in-Publication Data
A catalogue record for this book is available from the British Library

ISBN: 978-1-910190-73-9 (paperback)
ISBN: 978-1-910190-74-6 (ebook)

Cover design adapted from Figure 4.1.

Contents

	Preface	v
1	Enteroviruses Future Karla Kirkegaard	1
2	*Enterovirus* Receptors and Entry Jacqueline D. Corry, Jeffrey M. Bergelson and Carolyn B. Coyne	7
3	Hijacking Host Functions for Translation and RNA Replication by Enteroviruses Sonia Maciejewski and Bert L. Semler	23
4	The Omics of Rhinoviruses Ann C. Palmenberg	51
5	Viral Population Dynamics and Sequence Space Gonzalo Moratorio and Marco Vignuzzi	69
6	*Enterovirus* Control of Cytoplasmic RNA Granules Richard E. Lloyd	93
7	The Autophagic Pathway and *Enterovirus* Infection William T. Jackson	113
8	The Lipid Blueprints of Replicating Viral Genomes Nihal Altan-Bonnet, Marianita Santiana and Olha Ilnytska	129
	Index	145

Current Books of Interest

DNA Tumour Viruses: Virology, Pathogenesis and Vaccines	2018
Pathogenic *Escherichia coli*: Evolution, Omics, Detection and Control	2018
Postgraduate Handbook: A Comprehensive Guide for PhD and Master's Students and their Supervisors	2018
Molecular Biology of Kinetoplastid Parasites	2018
Bacterial Evasion of the Host Immune System	2017
Illustrated Dictionary of Parasitology in the Post-genomic Era	2017
Next-generation Sequencing and Bioinformatics for Plant Science	2017
The CRISPR/Cas System: Emerging Technology and Application	2017
Brewing Microbiology: Current Research, Omics and Microbial Ecology	2017
Metagenomics: Current Advances and Emerging Concepts	2017
Bacillus: Cellular and Molecular Biology (Third Edition)	2017
Cyanobacteria: Omics and Manipulation	2017
Foot-and-Mouth Disease Virus: Current Research and Emerging Trends	2017
Brain-eating Amoebae: Biology and Pathogenesis of *Naegleria fowleri*	2016
Staphylococcus: Genetics and Physiology	2016
Chloroplasts: Current Research and Future Trends	2016
Microbial Biodegradation: From Omics to Function and Application	2016
Influenza: Current Research	2016
MALDI-TOF Mass Spectrometry in Microbiology	2016
Aspergillus and *Penicillium* in the Post-genomic Era	2016
The Bacteriocins: Current Knowledge and Future Prospects	2016
Omics in Plant Disease Resistance	2016
Acidophiles: Life in Extremely Acidic Environments	2016
Climate Change and Microbial Ecology: Current Research and Future Trends	2016
Biofilms in Bioremediation: Current Research and Emerging Technologies	2016
Microalgae: Current Research and Applications	2016
Gas Plasma Sterilization in Microbiology: Theory, Applications, Pitfalls and New Perspectives	2016
Virus Evolution: Current Research and Future Directions	2016
Arboviruses: Molecular Biology, Evolution and Control	2016
Shigella: Molecular and Cellular Biology	2016
Aquatic Biofilms: Ecology, Water Quality and Wastewater Treatment	2016
Alphaviruses: Current Biology	2016
Thermophilic Microorganisms	2015
Flow Cytometry in Microbiology: Technology and Applications	2015

Full details at www.caister.com

Preface

The 12 species of the Enteroviruses – enterovirus A–H, enterovirus J, and rhinovirus A–C – are responsible, by many accounts, for more morbidity than any other viruses. The diversity of diseases caused by these genetically similar viruses is enormous, from the common cold to hand, foot and mouth disease to more serious diseases including cardiac infection, bulbar paralysis, and encephalitis.

Despite, or possibly because of, their success as pathogens, these prevalent and successful viruses function as highly efficient machines. Their entire genomes are usually under 8000 nucleosides, perhaps the size of two human genes, in a single positive-sense RNA molecule. The single open reading frames typically encode a single polyprotein which is produced in the absence of typical 5′ cap signals, through use of an internal ribosome entry site. The polyprotein is cleaved, by viral proteases encoded within the polyprotein itself, into the proteins required to facilitate virus replication. A subset of these proteins produce a negative sense copy of the genome, which in turn are used to template more positive sense genomes for further translation and, ultimately, packaging in nascent virions. The typical end of the cycle is cell lysis and virus release, although not all infections are lytic and virus can be shed throughout the life cycle as naked and enveloped virions.

Poliovirus, which remains by far the best-studied member of the genus, is on the verge of eradication in the wild. Yet much more work remains to be done, as improvements in available poliovirus vaccines will be needed to complete the final challenges of eradication. In the meantime, enteroviruses D68 and 71 have emerged as significant public health threats over the last decade, and while the available data from study of other Enteroviruses have jump-started research on these viruses, there are clearly enormous differences between the Enteroviruses, such that nothing can be taken for granted or assumed when studying a new member of the genus.

In this volume, some of the best researchers in the *Enterovirus* field take the reader on a tour of the most exciting frontiers in the study of the genus. From understanding viral entry into cells, translation of the genome, and RNA–RNA replication, to the dynamic genomics of these viruses, to studies of viral avoidance of host cell defenses and lipid-mediated exit from cells, the topics are cutting-edge and the expertise second to none.

We are proud to bring you a collection of chapters representing the best of the field and our latest understanding of the genus *Enterovirus*. We hope you enjoy reading it as much as we have enjoyed assembling and editing it.

William T. Jackson
Department of Microbiology and Immunology, University of Maryland School of Medicine, Baltimore, MD, USA

Carolyn B. Coyne
Department of Pediatrics, University of Pittsburgh, Pittsburgh, PA, USA

Enteroviruses Future

1

Karla Kirkegaard

Department of Genetics, Stanford University School of Medicine, Stanford, CA, USA.

Correspondence: karlak@stanford.edu

https://doi.org/10.21775/9781910190739.01

Abstract

At a time when poliovirus, the flagship *Enterovirus*, is on the brink of eradication, and while other Enteroviruses are simultaneously just emerging as serious public health threats, studies focused on the Enteroviruses and other picornaviruses are as important as ever. The true nature of cell exit and cell-to-cell movement by these viruses, for example, is only now being elucidated, particularly taking advantage of recent advances in cell modelling of physiologically relevant cell systems. Modern genomic techniques are just beginning to allow a population-level understanding of mutation and adaptation in these viruses, and are sure to reveal novel drug targets based on the *cis/trans* genetics of viral genomic regions. Continued understanding of the basic life cycle of these viruses, and of their genomes, will allow novel avenues of vaccine development. In the slightly more distant future, infection by Enteroviruses will be rapidly diagnosed and treatments may be tailored by personalized medicine. Finally, Enteroviruses are only beginning to be used as tools, particularly as anti-cancer therapeutics. It is impossible to see the future, but as the field moves forward, it is clear that both basic and applied *Enterovirus* research will remain an important topic for decades to come.

Introduction

It is a privilege to write an introduction to this compendium of cutting-edge *Enterovirus* research. We find ourselves at an interesting time. Poliovirus, we hope, is on the brink of eradication, due to the dedicated use of two very effective vaccines, although each has its drawbacks. Newly prominent pathogens such as enterovirus 71 and enterovirus D68 are currently causing considerable morbidity and mortality in humans. Foot-and-mouth disease virus is somewhat controlled, thanks to the world's first genetically engineered vaccine, but continues to be a dreaded scourge of livestock because of the rapid spread of even occasional outbreaks. An excellent subunit vaccine for hepatitis A has been developed, although its use in areas where hepatitis A is most threatening to human health is limited. No vaccines exist for rhinoviruses, coxsackieviruses, or any other Enteroviruses other than those mentioned above. No FDA-approved antiviral to any *Enterovirus* is available. Thus, intensive study of Enteroviruses is certainly warranted from the point of view of amelioration of human and animal disease. Both anticipated developments in vaccination and antivirals will be discussed below.

From a basic science perspective, we can certainly point to many findings – receptor-dependent tropism, internal ribosomal entry, RNA-dependent RNA polymerase activity and protease-mediated inhibition of critical cellular proteins, just to mention a few – that have proved ground-breaking in virology and beyond. However, given the spectacular and effective focus on hepatitis C virology in recent years, previous claims that Enteroviruses are the best-studied model systems for positive-strand RNA viruses have received a serious challenge. Yet, so many interesting questions remain!

Curiouser and curiouser

One of my favourite memories of any virus meeting is when Vadim Agol came slowly to the podium, appraised the audience and darkly intoned, 'There are many ways to die'. This Dostoyevskian preface segued into a discussion of poliovirus's ability to inhibit early innate immune responses, prolonging the life of an infected cell, only to let it die later after the virus has had a chance to replicate. It has long been a source of anxiety to me that we did not know exactly why cells infected with any *Enterovirus* lived or died. Now, with the proliferation of cell death mechanisms – apoptotic, necrotic, pyroptotic, necroptotic and autophagic, in which the same signal can also lead to different outcomes in different cell types, and the same cell type under different circumstances – this seems less embarrassing. We are understanding more and more about the tissues in which viruses replicate, those through which they spread, and those in which they cause disease. Given that these tissues can be very different, much about viral strategy can be learned in the contemplation of which cells live and which cells die. The balance of life and death, so important to understanding how viruses spread through tissues, is likely to be influenced by everything we know about the cell biology of infection – from which cells show the strongest inhibition of translation to which cells are polarized and allow viral egress only directionally. Pharmaceuticals against specific kinds of cell death, such as TNF inhibitors and necrostatins, are currently available and more are promised due to intensive research on tissue-sparing treatments for neurodegenerative disease.

Darwinian evolution requires pre-existing diversity. Then, successful genomes are selected from that diverse pool. Thus, the initial postulates of 'adaptive mutation', in which the exertion of selective pressures could increase mutation rate, at first seemed frighteningly Lamarckian. Could selection pressure itself actually create mutations? Yes, it turns out, because the exertion of selective pressure can induce stress responses in many bacteria. Part of the bacterial stress response is an increase in mutation frequency, which will then increase the likelihood that some individuals within the threatened population will survive the selective pressure. Could the concept of adaptive mutation be relevant to *Enterovirus* infection? The incredible power of deep sequencing approaches will continue to elucidate this and other fascinating transmission genetic problems in *Enterovirus* biology.

Whether a virus population can survive selection pressure, such as spread to a different tissue in the same host, in the presence of an antiviral compound or increasing concentrations of antibodies depends on many factors. First, there is the intrinsic mutation rate based on polymerase fidelity. We do not actually know this number for any *Enterovirus* polymerase in the context of an infected cell. What is measured in the many elegant experiments to define the intracellular quasispecies is the cumulative mutation frequency after several intracellular cycles of RNA synthesis, and the structure of the intracellular generations will greatly influence the cumulative mutation rate. For example, if each positive strand generated one negative strand, and each negative strand templated 50 positive strands, only six templated replicative events, or RNA generations, would be required to generate more than 1000 intracellular positive strands. Thus, the cumulative error rate will be six times the intrinsic error rate. If, on the other hand, each negative strand generated only five positive strands, ten RNA generations would be required to generate more than 1000 intracellular positive strands. It is therefore predicted that selection pressures that inhibit positive-strand synthesis might actually increase the cumulative error rate, a form of adaptive mutation that will be interesting to test and could be important for cell-to-cell spread and response to treatments.

This also brings up the general question whether errors and recombination events occur with fixed probability in a Poisson distribution. It is possible that there are RNA replication complexes that are more recombinogenic than others. It will be very interesting, now that single-molecule sequencing is available and error-proofed to *determine the error rates and recombination frequencies of individual RNA replication complexes*. Perhaps RNA replication complexes associated with different host factors or assembled on different organelles, or with different ratios of processed polymerase to catalytically inactive precursors, will manifest different amounts of fidelity or processivity. It would make sense, evolutionarily, to have some sober and some deranged RNA replication complexes. When attending a party, for example, it is often a good idea to bring a boisterous friend. Her antics will increase your contacts and allow your inclusion in after-party activities but, the next day, you do not actually have to be like her. In short, as our sequencing abilities become more and more focused, we are likely to find several levels to viral diversity.

Viral eradication and control by vaccination

Thus far in human history, three viruses have been eradicated: smallpox (in 1978), poliovirus serotype 2 (in 1999) and rinderpest (in 2011). We all fervently hope that the hard-fought campaign to eradicate the remaining two strains of poliovirus will be successful, and soon. As of this writing, the only established endemicity is of serotype 1 poliovirus in Pakistan, Afghanistan and Nigeria. However, wild-type strains of both serotypes 1 and 3 continue to circulate in several countries, including recent environmental sampling in Israel. To the extent that basic science informs and interprets the current eradication campaign, it is interesting to consider what we can learn from this in the management and treatment of other human maladies.

The complexities of our reliance on the Sabin vaccine to prevent poliovirus was made especially explicit by experiments of Phil Minor and his colleagues, beginning in the 1980s, in the 'nappies' of healthy children, including offspring, were monitored immediately after administration of the trivalent Sabin vaccine. Within one day, it was possible to observe selection for nucleotide changes that conferred neurovirulence. This finding rationalized what had been discovered empirically during the vaccination campaign: that use of live attenuated massive vaccination coverage was needed to ensure that everyone was actually vaccinated with attenuated virus, rather than second hand from a vaccinee. Without the availability of infectious cDNA clones of types 1, 2 and 3 poliovirus, and the mouse models with which to correlate the neurovirulence of mutations, it would likely have been difficult to test any causation relationships of individual mutations identified by epidemiological surveillance.

We can bring many of our most interesting molecular genetic strategies to bear on the design of live attenuated viruses that will not be vulnerable to pathogenic reversion in the vaccinee. A particularly innovative solution is the 'death by 1000 cuts' of codon de-optimization. Beginning with poliovirus and then extending to other infectious agents such as influenza, genomes can be re-coded to increase the representation of infrequent codon pairs. The amount of viral attenuation conferred in this way related almost linearly to the length of sequence so altered, and adaptation to improved growth characteristics was not observed. The molecular basis for this useful strategy first revealed by the laboratory of Olin Kew is that the unpopular codon pairs contained CpG and UpA dinucleotides, thus targeting the RNAs for destruction by the innate immune response. Thus, *attenuation of vaccine strains by recoding to introduce more CpG and UpA dinucleotides may provide a much-needed recipe for live vaccine development.* Many fascinating questions remain. The known recognition of CpG dinucleotides by the intra-endosomal TLRs 9 and 21 may be responsible for some of this effect, but there are likely to be aspects of the innate immune response revealed by these observations that we do not understand. Are there particular regions of the host cells studied thus far in which this restriction occurs? Which viruses, and which RNAs of those viruses, are the most vulnerable to these destructive mechanisms? Looking forward, it will be interesting to learn whether all hosts, and all tissues within those hosts, exhibit the innate immune response that leads to RNA modification and destruction at CpG and UpA dinucleotides. It is very possible that the observed restriction is general among mammals, given that many cellular mRNAs have undergone a similar selection pressure.

Antivirals: crucial for post eradication of poliovirus and needed for all Enteroviruses

After poliovirus is eradicated, vaccination will cease. It will require a combination of high previous vaccine coverage and very good luck for the recently vaccinated and chronic excreters not to serve as potent reservoirs of potentially neurovirulent virus. The realization that this problem could compromise the goal of poliovirus eradication has been discussed at length, beginning with a NIH panel 'The Role of Antivirals in Poliovirus Eradication' first convened in 2006. At that time, inhibitors of capsid function and inhibitors of the major *Enterovirus* protease, 3C, were considered the leading candidates for antivirals to be used in such an outbreak.

Since that time, poliovirus capsid inhibitors have progressed into clinical trials, which will also reveal whether or not drug resistance is associated with monotherapy. My laboratory has shown that the use of capsid inhibitors is not associated with high frequencies of drug resistance due to the intracellular dominance of the drug-susceptible capsids. The availability of such treatments such treatment could go a long way towards ameliorating post-eradication outbreak and supporting public acceptance of large-scale vaccination and the colossal effort required to eradicate a major human pathogen. From this effort, we are likely to learn whether monotherapies are possible if care is used to suppress the outgrowth of drug resistance, whether infected individuals can be cured if identified sufficiently early and whether antiviral use, vaccination or both should be used to control local flare-ups of infection. These principles will be important in the control of other *Enteroviruses* in response to natural infections and to provide a safety net for the medical use of *Enteroviruses* as vaccine vectors and cancer treatments.

Enteroviruses in 10 years

If *Enterovirus* research is allowed to continue at the rate it has for the last decades, tremendous contributions will be made to basic and clinical science. In 10 years, we will have used single-molecule tracking to watch enteroviral capsids ratchet open along their most vulnerable symmetry axes to release viral RNA. We will have seen RNA templates and newly synthesized RNA speeding along replicative lattices, and used differential high-resolution fluorescence to reveal the local protein and lipid compositions for positive-strand and negative-strand synthesis. We will have learned why dendritic cells can be inactivated by enteroviral infection and thus developed strategies to maintain these crucial cells for all infections. The elegance of the *Enterovirus* capsid, our deep understanding of the human immune response to oral infection, and the wide availability of inexpensive capsid inhibitors for all Enteroviruses will make them ideal vaccine vectors for much more complicated entities. Most exciting is the targeting of Enteroviruses for the treatments of cancers. For these applications, the plethera of viruses and serotypes will be a blessing, because they can serve as the basis for multiple sequential vaccines and treatments.

Allowing one's imagination further latitude, in ten years we will have interrogated the mouse Collaborative Cross and the Human 1000 Genomes Project (perhaps 100,000 genomes by then!) by infection with multiple Enteroviruses. We will have delineated the alleles that render us more or less likely to support viral growth, virally induced inflammation and personal morbidity. When we visit with a fever, the doctor will quickly ascertain whether we have an enteroviral infection using a dipstick ELISA. The recommendation could be 'Your lipid profile has always shown a high and variable abundance of PI4P, and it is especially elevated now. In the short term, I'd like to put you on this PI4KIIIβ inhibitor, which should be active against any positive-strand RNA virus, and run a genechip tomorrow so we can chose the capsid inhibitor specific for your virus'. Or, the doctor might say 'Well, you're practically null for RPS25 expression, which is why you rarely get these infections. Go home, eat well, take two amino acid supplements to suppress autophagy, and call me in the morning'. A continued challenge will be to ensure that effective care can be extended to all the world's citizens, given that enteroviral infections can preferentially affect the relatively disenfranchised. This consideration further emphasizes the need for public or philanthropic funding of basic and clinical research in *Enterovirus* research and in other areas that deeply affect human health, well-being, and opportunity.

Enterovirus Receptors and Entry

2

Jacqueline D. Corry[1], Jeffrey M. Bergelson[2] and Carolyn B. Coyne[1]*

[1]Department of Pediatrics, University of Pittsburgh, Pittsburgh, PA, USA.
[2]Division of Infectious Diseases, Children's Hospital of Philadelphia, PA, USA.
[3]Department of Pediatrics, University of Pennsylvania, Philadelphia, PA, USA.

Correspondence: coynec2@pitt.edu

https://doi.org/10.21775/9781910190739.02

Abstract

In order to invade the host, Enteroviruses must first attach to receptors expressed on the cell surface and undergo a series of events that culminate in genome release. However, *Enterovirus* receptors serve functions beyond that of mere docking sites and it is now clear that intracellular signals initiated by receptor binding prime the host cell for virus internalization by promoting modification(s) of the host cell that facilitates endocytic uptake. For many Enteroviruses, these processes are complicated by the inaccessibility of their receptors to cellular junctions and/or to the complex environment in which they are interacting with their target cells, such as the gastrointestinal tract. In this chapter, we discuss the diverse cellular factors that serve as *Enterovirus* receptors, the steps involved in *Enterovirus* genome release, and the endocytic pathways utilized by these viruses to gain access to the host cell cytosol.

Introduction

An *Enterovirus* consists of a viral RNA genome encased within a protective protein shell. Viral RNA is infectious when introduced into cells. Thus, *Enterovirus* entry is, in essence, the process by which viral genomes are delivered from the external environment to the cell cytoplasm where replication occurs. Entry begins with attachment of the virion to the cell surface, most often to a specific receptor molecule; in subsequent events the virion is internalized within an endocytic vesicle, RNA is released from the capsid (in a process referred to as 'uncoating'), and free RNA is delivered across a vesicular membrane into the cytoplasm. Like most other events in the viral life cycle, viral entry makes use of a variety of the host cell's endogenous mechanisms – most notably, mechanisms for endocytosis and vesicular transport – and depends on a large number of host molecules

A variety of cell surface receptors facilitate *Enterovirus* entry and uncoating, and Enteroviruses use a variety of endocytic routes to reach the cytoplasm. In this chapter, we review the general properties of the *Enterovirus* capsid and mechanisms for genome release, discuss the cell surface receptors used to facilitate *Enterovirus* attachment, and outline the strategies used by these viruses to gain entry into the cell cytoplasm. Although *human rhinoviruses* have recently been assigned to the genus *Enteroviridae,* their instability at low pH distinguishes them from other Enteroviruses such as polioviruses, coxsackieviruses, echoviruses, and enterovirus 71, which has an effect on mechanisms of entry, which we will address separately.

The *Enterovirus* capsid

The *Enterovirus* capsid is an icosahedron made up of 12 pentamers, with each pentamer containing five protomers, each composed of four capsid proteins (VP1–VP4) (Hogle *et al.*, 1985; Rossmann *et al.*, 1985). VP1–3 are exposed on the capsid surface, whereas VP4, a small myristoylated protein (Chow *et al.*, 1987), is entirely internal (Hogle *et al.*, 1985; Rossmann *et al.*, 1985). In many Enteroviruses (and Rhinoviruses), a deep depression, referred to as the 'canyon' surrounds each 5-fold axis of symmetry (Filman *et al.*, 1989; Muckelbauer *et al.*, 1995; Rossmann *et al.*, 1985); a number of *Enterovirus* receptors – including the receptors for polioviruses (Belnap *et al.*, 2000b), many rhinovirus serotypes (Olson *et al.*, 1993; Kolatkar *et al.*, 1999), and group B coxsackieviruses (He *et al.*, 2001)— are known to bind within the canyon. In contrast, several other enterovirus and rhinovirus receptors bind outside the canyon (He *et al.*, 2002; Hewat *et al.*, 2000).

It has been suggested that, within the canyon, the receptor binding site is inaccessible to neutralizing antibodies, and shielded from immune pressures that drive antigenic variation in more exposed regions of the capsid (Rossmann, 1989). This 'canyon hypothesis' may explain how more than 90 antigenically distinct rhinovirus serotypes maintain the capacity to bind a single receptor (Rossmann *et al.*, 1985; Colonno *et al.*, 1988; Uncapher *et al.*, 1991). However, despite the hypothesis, it is now clear that some neutralizing antibodies are capable of penetrating the canyon (Smith *et al.*, 1996; Chen *et al.*, 2013).

Beneath the canyon floor is a small hydrophobic pocket, connected by a pore to the canyon, and filled by a low molecular weight molecule (the 'pocket factor', most likely a fatty acid) (Hogle *et al.*, 1985; Rossmann *et al.*, 1985). The pocket factor is thought to stabilize the virion. Antiviral compounds such as pleconaril bind within the pocket, displacing the pocket factor, and inhibit infection by preventing the capsid from undergoing conformational changes associated with uncoating and RNA release (Grant *et al.*, 1994; Rogers *et al.*, 1999).

Attachment to a cellular receptor

The first step in picornavirus entry is attachment to a receptor molecule on the cell surface. Receptors concentrate virions at the cell surface, increasing the likelihood that infection will occur, and in many cases, they also facilitate infection by other mechanisms. Receptor contact may initiate the uncoating process, and endocytosis of a receptor-bound virion may facilitate virus delivery to an intracellular compartment where RNA release occurs. In some cases, virus receptors also transduce intracellular signals that promote entry and infection.

In the late 1950s and early 1960s, it was found that homogenates of primate cells or tissues susceptible to infection by *poliovirus* were capable of adsorbing poliovirus (PV) virions from solution (McLaren *et al.*, 1959, Holland, 1961); in contrast, homogenates obtained from non-susceptible rodent or rabbit cells, did not adsorb virus, suggesting that primate cells expressed specific receptors capable of binding virus. However, when exposed to isolated viral RNA, non-primate cells as well as primate cells became productively infected (Holland *et al.*, 1959). PV virions were found to bind to homogenates of monkey tissues only from those organs known to be targeted by PV *in vivo*. Experiments like these led to the idea that specific receptors are important factors in determining the tropism of PV and other *Enteroviruses* for specific species and tissues. The receptors for PV and *coxsackievirus* (CV) B were soon found to associate with the membrane fraction of cell homogenates and to be susceptible to digestion by proteases (Holland, 1961), but the specific cell surface plasma membrane proteins that serve as receptors for PV (Mendelsohn *et al.*, 1989) and CVB (Bergelson *et al.*, 1997) were not identified for 30 years.

Receptors have now been identified for a number of Enteroviruses (Table 2.1 and associated references). The poliovirus receptor (PVR), the receptor for group B coxsackieviruses (the coxsackievirus and adenovirus receptor, CAR), and the receptor for nearly 90 human rhinoviruses (intercellular adhesion molecule-1, ICAM-1) are comprised of multiple immunoglobulin-like domains; in each case, the N-terminal domain of the receptor inserts into the viral canyon (Belnap *et al.*, 2000b; He *et al.*, 2001; Olson *et al.*, 1993; Xing *et al.*, 2000; He *et al.*, 2000). The receptor for echovirus 1 is an integrin molecule, as are receptors identified for parechoviruses and some group A coxsackieviruses; integrin α2β1 (specifically, the α2 I- domain) (King *et al.*, 1995) binds within the echovirus 1 canyon (Xing *et al.*, 2003), whereas αv integrins bind to peptide loops, containing the integrin recognition sequence Arg-Gly-Asp (RGD), exposed on the capsids of parechoviruses (Boonyakiat *et al.*, 2001) and coxsackie A viruses (Roivainen *et al.*, 1991). Decay accelerating factor (DAF), a complement regulatory protein that serves as a receptor for multiple echoviruses and coxsackieviruses, binds outside the viral canyon, draping across the capsid surface (He *et al.*, 2002; Hafenstein *et al.*, 2007); it is interesting that different viruses appear to interact with different areas of the DAF molecule, suggesting that they may have evolved independently to bind the same receptor (Powell *et al.*, 1999). Although most of the identified receptors are proteins, a number of Enteroviruses have also been shown to bind to carbohydrate moieties; interaction with negatively-charged polysaccharides (such as heparan sulfate) in several cases depends on the presence of acidic amino acid residues clustered at the 5-fold axis of the viral capsid (McLeish *et al.*, 2012).

Use of multiple receptors

A number of Enteroviruses bind to more than one receptor. Using multiple receptors may permit a virus to infect a broader range of cell types, but interaction with multiple receptors on a single cell may be required for a virus to complete particular steps in the entry process. The use of multiple receptors by a single virus was first described for coxsackie B viruses (CBV) (Hsu *et al.*, 1988); all CBV bind to CAR (Bergelson *et al.*, 1997; Tomko *et al.*, 1997), but a subset of CBV also bind to DAF (Bergelson *et al.*; 1995, Shafren *et al.*, 1995). Whereas virus attachment to CAR initiates the uncoating process, attachment to DAF does not (Milstone *et al.*, 2005); on cells that express both CAR and DAF, virus interaction with DAF promotes infection by enhancing virus attachment, but after binding to DAF, viruses must

Table 2.1 Viral entry factors

Virus	Receptor
Poliovirus	PVR (CD155) (Mendelsohn *et al.*, 1989)
Rhinovirus (major group)	ICAM-1 (Greve *et al.*, 1989, Staunton *et al.*, 1989, Tomassini *et al.*, 1989)
Rhinovirus (minor group)	LDL receptor (Hofer *et al.*, 1994)
Rhinovirus C	CDHR3 (Bochkov *et al.*, 2015)
Coxsackie B viruses	CAR (Bergelson *et al.*, 1997, Tomko *et al.*, 1997)
	Decay accelerating factor (DAF, CD55) (Bergelson *et al.*, 1995, Shafren *et al.*, 1995)
	Heparan sulfate (Zautner *et al.*, 2003)
Echovirus 1	Integrin $\alpha2\beta1$ (Bergelson *et al.*, 1992)
Other echovirus serotypes	DAF (CD55) (Bergelson *et al.*, 1994, Ward *et al.*, 1994, Powell *et al.*, 1998)
	Heparan sulfate (Goodfellow *et al.*, 2001)
	Integrin $\alpha v\beta3$ (Ylipaasto *et al.*, 2010)
	Evolution of viruses that bind DAF (Powell *et al.*, 1997)
Parechoviruses	Integrins $\alpha v\beta3$ (Triantafilou *et al.*, 2000), $\alpha v\beta1$(Triantafilou *et al.*, 2000), $\alpha v\beta6$_(Seitsonen *et al.*, 2010)
	Heparan sulfate (Merilahti *et al.*, 2016)
Coxsackie A viruses	ICAM (Shafren *et al.*, 1997a)
	DAF (Shafren *et al.*, 1997b)
	Integrin $\alpha v\beta3$ (Roivainen *et al.*, 1994), $\alpha v\beta6$ (Heikkila *et al.*, 2009)
	MHC-I -associated GRP78 (Triantafilou *et al.*, 2002)
	SCARB2 (Yamayoshi *et al.*, 2009, Yamayoshi *et al.*, 2012)
	Sialic acid (Nilsson *et al.*, 2008)
	Heparan sulfate (Merilahti *et al.*, 2016)
	PSGL-1 (Nishimura *et al.*, 2009)
Enterovirus 68	Sialic acid (Imamura *et al.*, 2014, Liu *et al.*, 2015)
	ICAM-5 (Wei *et al.*, 2016)
Enterovirus 70	DAF (Karnauchow *et al.*, 1996)
	Sialic acid (Alexander and Dimock, 2002)
Enterovirus 71	PSGL-1 (Nishimura *et al.*, 2009)
	SCARB2 (Yamayoshi *et al.*, 2009)
	Heparan sulfate (Tan *et al.*, 2013)
	Vimentin (Du *et al.*, 2014)
	Nucleolin (Su *et al.*, 2015)
	Annexin II (Yang *et al.*, 2011)

also interact with CAR for uncoating to begin. This separation of functions is particularly clear in polarized epithelial cells (such as those that line the intestine), where CAR and DAF are located in different places – DAF on the apical cell surface, and CAR in intercellular tight junctions (Shieh and Bergelson, 2002). Viruses bind initially to DAF on the apical cell surface, then move to tight junctions, where contact with CAR initiates uncoating and subsequent events in entry (Coyne and Bergelson, 2006). As is discussed in more detail below, virus attachment to DAF on the apical surface also leads to transmission of intracellular signals, important both for virus movements across the cell surface and for uptake of virions from tight junctions into the cell (Coyne and Bergelson, 2006).

Uncoating: formation of expanded A-particles

For many Enteroviruses, uncoating is initiated by binding of a receptor within the canyon, with formation of an altered particle (or A-particle). A-particles (and in some cases, 80S empty capsids) are formed when virions are incubated *in vitro* with isolated receptor protein at 37 °C (Yafal *et al.*, 1993; Xing *et al.*, 2000; Milstone *et al.*, 2005). Deformation at the base of the canyon leads to expulsion of the pocket factor (Butan *et al.*, 2014), exposure of the hydrophobic N-terminus of VP1 at the capsid surface (Fricks and Hogle, 1990), and release of VP4 from the virion (Crowell and Philipson, 1971). A-particles are somewhat larger than native virions [expanded in diameter by about 4% (Belnap *et al.*, 2000a)], and can be distinguished from native virions by their decreased sedimentation velocity in sucrose gradients (135S, as opposed to 160S for native virions or 80S for empty capsids from which RNA has been released) (Crowell and Philipson, 1971). Once they are extruded from the capsid, the VP1 N-terminus and VP4 associate with the cell membrane through hydrophobic interactions. VP1 anchors the virion to the membrane, and both VP1 and VP4 promote the formation of pores (Danthi *et al.*, 2003; Tosteson and Chow, 1997) through which viral RNA is believed to enter the cell (Groppelli *et al.*; 2017, Danthi *et al.*, 2003; Moscufo *et al.*, 1993).

In general, receptors that bind outside the canyon do not trigger A-particle formation, and viruses that bind these receptors must require another trigger for uncoating. In some instances, the triggering event is interaction with a second receptor, as described above for CVB interactions with DAF and CAR. This is also the case for enterovirus 71 isolates that bind to the leucocyte molecule P-selectin glycoprotein ligand (PSGL)-1. PSGL-1, which appears to interact with a plateau at the 5-fold axis (Nishimura *et al.*, 2015, Nishimura *et al.*, 2013), does not trigger formation of A-particles (Yamayoshi *et al.*, 2013); in contrast, A-particles are formed when EV71 interacts with the lysosomal protein SCARB2 (scavenger receptor class B, member 2) (Yamayoshi *et al.*, 2013).

A number of rhinoviruses undergo conversion to A-particles when they are exposed to acid, even in the absence of receptor contact (Korant *et al.*, 1972; Nurani *et al.*, 2003). For those viruses that use the low-density lipoprotein receptor (LDLR, which binds at the 5-fold vertex) for cell entry, the primary function of the receptor is to deliver the virion to an acidic endosome. Among the many rhinoviruses that bind ICAM-1, uncoating of some viruses is initiated on contact with the receptor, but for others, endosomal acidification is also required (Nurani *et al.*, 2003).

Uncoating: RNA release

Expansion of the capsid opens gaps between protomers near the 2-fold axes, as well as smaller gaps at the base of the canyon (through which the VP1 N-terminus protrudes) (Garriga *et al.*, 2012; Ren *et al.*, 2013; Butan *et al.*, 2014). Cryo-electron microscopy studies of heated poliovirus particles 'caught' while undergoing RNA release reveals RNA density emerging near the 2-fold axes, suggesting that it may exit through openings at the 2-fold axes (Bostina *et al.*, 2011). Work with acid-treated rhinovirus indicates that RNA release occurs in an ordered fashion, with the 3′-poly-A tail emerging from the capsid first (Harutyunyan *et al.*, 2014; Harutyunyan *et al.*, 2013). Release of the RNA leaves behind an empty capsid shell, which can be detected in infected cells as an 80S viral particle.

Although heat treatment and acid treatment induce global changes in the capsid structure, virus attachment to cells is asymmetric, beginning with attachment to a single receptor, and efficient infection would require that RNA release be directed towards the membrane attachment site and across the membrane. Cryo-EM studies of virions bound to receptors immobilized on lipid membranes have recently begun to reveal some of the more subtle, asymmetric conformational changes that lead to RNA release *in vivo*. When CVB3 is bound in the cold to a single CAR molecule immobilized on a lipid bilayer nanodisc, brief exposure to physiologic temperature (37 °C) results in formation of a partially expanded particle, with conformational changes – loss of pocket factor and extrusion of VP1 – largely restricted to the site of receptor contact (Lee *et al.*, 2016); a local disruption of the capsid surface may provide an exit site for RNA. It is unknown whether RNA release *in vivo* occurs from such an asymmetric uncoating intermediate, or whether RNA is released after from a more classic A-particle after local conformational changes have been propagated through the capsid.

Where does uncoating occur?

For many Enteroviruses (as opposed to rhinoviruses), A-particle formation may begin as soon as a virion contacts a canyon-binding receptor at the cell surface. CBV and poliovirus A-particles appear shortly after infection, even when endocytosis of virions has been blocked with chemical inhibitors (Coyne and Bergelson, 2006; Coyne *et al.*, 2007). However, RNA release takes longer, and likely does not occur until the virion has been internalized in an endocytic vessel (Brandenburg *et al.*, 2007; Coyne and Bergelson, 2006). The trigger for RNA release within endocytic vesicles has not been identified. Although initial reports suggested that endosomal acidification might be required for poliovirus entry (Madshus *et al.*, 1984), subsequent work revealed that uncoating could proceed even when acidification was blocked by bafilomycin, a powerful inhibitor of the endosomal proton pump (Pérez and Carrasco, 1993). Recent work indicates that a cellular phospholipase, PLA2G16 is important for RNA release into the cytoplasm (Staring *et al.*, 2017) by several Enteroviruses. PLA2G16 appears to localize to virus-containing endosomes in response to virus-induced pore formation, but its precise mechanism of action remains uncertain.

Introduction to endocytosis

Non-enveloped viruses gain entry into cells through diverse pathways, with the route often directly influenced by cell type and receptor expression or localization. Many studies have interrogated the pathways utilized by Enteroviruses to enter host cells. Below, we provide a brief overview of the types of endocytosis and then highlight key studies related to specific members of the *Enterovirus* family.

Clathrin-mediated endocytosis

Clathrin-mediated endocytosis (CME) occurs within clathrin-coated pits and vesicles, components of which include clathrin heavy and light chains and the adaptor protein 2 (AP-2) complex. Clathrin-coated pits and vesicles are small (≈100 nm), uniform in appearance, and visualized by their electron dense clathrin coats by transmission electron microscopy. Following pit formation, clathrin-coated vesicles require the activity of the dynamin GTPase, which undergoes hydrolysis to catalyse membrane fission (Elkin *et al.*, 2016) and subsequent entry into the cell cytoplasm.

Caveolar endocytosis

Caveolar endocytosis is reliant on caveolins, cavins (cytosolic coat proteins), cholesterol, sphingolipids, and lipid rafts. Caveolae are caveolin-coated pits that are also involved in cellular signalling, lipid metabolism, and surface-tension sensing. Caveolin-1 is an integral membrane protein that inserts into the membrane and binds to cholesterol to serve as a scaffold for subsequent signalling molecules. Caveolin-mediated endocytosis is triggered by ligand binding to receptors concentrated within lipid rafts or caveolae that have formed on the cell surface. Several kinases and phosphatases are required for caveolar budding and as with clathrin-coated pits, dynamin activity is required for fission of caveolin-coated vesicles (Elkin *et al.*, 2016; Mayor *et al.*, 2014).

Macropinocytosis

Macropinocytosis is the bulk, non-selective uptake of nutrients, antigens, and other extracellular molecules. While this process is constitutively active in macrophages, it is not active in other cell types and must first be stimulated. Macropinocytosis involves membrane ruffling, which requires massive rearrangements of the actin cytoskeleton, which require the activity of a number of components including PI3-kinase, small GTPase Rac1 and Cdc42, amongst others. In addition to the requirement to become activated, macroponocytosis can be distinguished from other pathways such as CME based upon the large size (> 1–2 μm in many cases) of internalized vesicles.

Other forms of endocytosis

There are other forms of endocytosis that are neither clathrin- nor caveolin-dependent, nor fall into the definition of bulk fluid phase uptake pathways like macropinocytosis. Although these pathways are sometimes referred to as 'alternative' pathways, these pathways may function as frequently as the 'classical' forms, but the lack of specific markers for these pathways precludes their complete characterization or identification. Like these classical pathways, 'alternative' entry routes require actin cytoskeletal reorganizations. For example, RhoA-dependent endocytosis requires actin rearrangements, cholesterol, lipid raft-like domains and requires dynamin for vesicle fission. This pathway is important for cytokine receptor endocytosis and is also regulated by signalling through Rac1 and p21-activated kinase 1 (Pak1) (Mayor *et al.*, 2014).

Coat- and dynamin-independent mechanisms of entry are less-well-understood.

The clathrin-independent carriers/GPI-enriched early endosomal compartments (CLIC/GEEC) pathway is involved in endocytosis of lipid-anchored proteins such as glycophosphatidylinositol-(GPI) anchored protein. This pathway is a high-capacity pathway that recycles a large fraction of the membrane, but is distinct from non-specific uptake pathways such as macropinocytosis, as it is not sensitive to agents such as amiloride, that are thought to specifically target Rac1 and Cdc42 activity. It is unclear whether this pathway is dependent on cholesterol or lipid rafts; however, it does require Arf1 signalling, which is used to maintain the cycling of Cdc42 to recruit the actin machinery, which are necessary for macropinocytosis (Mayor *et al.*, 2014).

Lastly, flotillins mediate fluid-phase uptake and endocytosis by this method requires that the non-receptor Src kinase family member Fyn phosphorylate flotillin. cAbl and c-Src tyrosine kinases must also be activated to activate the Rac1 GTPase required for actin polymerization. Flotillin-mediated endocytosis may be either dynamin-independent or dynamin-dependent (Mayor *et al.*, 2014).

Post-internalization events

With rare exceptions, all endocytosed vesicles are directed to and fuse with a sorting endosome post-internalization. The pH of the sorting endosome is 5.5–6 (Müller *et al.*, 2012; Maxfield and McGraw, 2004), a pH at which ligands often dissociate from their associated receptors (Müller *et al.*, 2012). Internalized viral particles can be trafficked to the Golgi complex, the recycling endosome, and/or the late endosome, although the ultimate fate of these particles is often mediated by the type of vesicle in which they are localized. From the Golgi complex, the virus could be trafficked to the endoplasmic reticulum (ER), the lysosome or back to the cell surface. From the recycling endosome, the virus can be taken to the plasma membrane or to the Golgi complex. Finally, a virus that is trafficked to the late endosome can either be taken to the Golgi or the lysosome. In the lysosome, the virus can remain, be returned to the late endosome, or be degraded (Maxfield and McGraw, 2004; Wandinger-Ness and Zerial, 2014).

Endosome sorting is energy-dependent, and as such, Rab-GTPases are critical players in these sorting pathways and events. The Rab proteins are on the cytoplasmic interface of membranes and rely on GTP/GDP cycling for assembly of machinery within the cell. The machinery is dynamic and regulated. Rab proteins undergo conformational changes upon GTP binding and hydrolysis that enable their association with other proteins (Wandinger-Ness and Zerial, 2014). Rab GTPases control distinct phases and aspects of endocytosis and vesicular trafficking—for example, Rab5 is important for the maturation of the early endosome, Rab7 is important for the transition of the early endosome to the late endosome and further to the lysosome (Wandinger-Ness and Zerial, 2014; Numrich and Ungermann, 2014), and Rab34 is a Golgi-bound Rab involved in regulation of macropinocytosis. There are 66 known Rab proteins encoded in the human genome, including isoforms that have distinct functions (Wandinger-Ness and Zerial, 2014).

Below, we highlight several examples of the routes utilized by select Enteroviruses to gain entry into host cells.

Poliovirus

Utilizing cutting edge microscopy and a dual labelling approach that labelled the PV capsid with one fluorophore (Cy5) and the vRNA with another (Syto82), Brandenburg *et al.* showed that in HeLa cells, PV entry was rapid, with viral particles entering and releasing vRNA as early as 10 minutes following initiation of entry. More than half of the viral particles had released their vRNA by 22 ± 3 minutes post-infection, supporting the very rapid kinetics of entry and uncoating (Brandenburg *et al.*, 2007). This approach was coupled with a neutral red assay, whereby vRNA is labelled with neutral red, a compound that crosslinks vRNA upon exposure to light (Crowther and Melnick, 1961; Brandenburg *et al.*, 2007). This assay also supported rapid kinetics of vRNA release within 27 ± 3 mins of infection, which could be prevented when the actin cytoskeleton was depolymerized with cytochalasin D. RNAi-based approaches were used to silence the expression of the clathrin-heavy chain protein, the clathrin associated protein AP2μ or flotillin, none of which were required for vRNA release. Taken together, this study suggested that the PVR-mediated uncoating of PV occurred at or very near the cell surface and did not require classical receptor-mediated endocytosis to facilitate genome release.

The above-described study provided fundamental and important insights into the mechanisms accompanying PV entry and uncoating in host cells. However, as these studies were performed in a non-polarized cell, it remained unclear whether or not the mechanisms used by the virus to facilitate genome release were shared in polarized cell types, which have distinct apical and basolateral domains separated by junctional complexes. Using human brain microvascular endothelial cells (HBMEC), a cell-based model of the blood–brain barrier (Stins *et al.*, 2001; Drummond *et al.*, 2016), others suggested that the kinetics of PV entry and genomes uncoating were far slower and required a cascade of signalling events that involved actin cytoskeletal reorganization and Rho GTPase expression (Coyne *et al.*, 2007). In addition, PV entry into polarized cells required components associated with endocytosis, such as caveolin and dynamin as well as with cholesterol. However, as this study did not directly label the vRNA, it is unclear at what stage of this process uncoating occurred. However, inhibiting these post-attachment events prevented subsequent viral replication, suggesting that vRNA release occurred at some point post-internalization (Coyne *et al.*, 2007).

Coxsackievirus

The mechanisms by which CVB enters non-polarized and polarized cells types have also been defined and suggest that the pathways utilized by Enteroviruses to enter these distinct cell types may be divergent. Studies utilizing Caco-2 cells, an intestinal epithelial cell line isolated from colorectal cancer, revealed that the entry of CVB3 into these cells is highly complex and involves myriad cellular signalling pathways and molecules. The pathways that facilitate viral entry included Src family tyrosine kinases, Abl tyrosine kinase, Rho GTPases, cavelaor components, micropinocytosis mediators, tight junction components, amongst others (Coyne and Bergelson, 2006). Interestingly, these signals pathways were initiated by the viral-induced clustering of the apically-localized DAF receptor, forming a 'signalsome' by virtue of the lipid-enriched membrane domain formed by the clustering of the DAF GPI anchor. In addition, the clustering of DAF also led to the relocalization of the receptor from the apical surface to the tight junction, where the virus was then in close proximity to CAR, which is required to initiate uncoating. Interestingly, all of these events could be mimicked by an antibody which also induced DAF clustering, suggesting that the virus evolved a strategy to co-opt a phenomenon associated with this process to facilitate its entry.

Perhaps not surprisingly, in non-polarized cells, the process is less complex. In HeLa cells, CVB3 entry does not require DAF-mediated signalling, likely given that in these cells, CAR is not inaccessible by virtue of its tight junction localization (Patel *et al.*, 2009). In addition, the endocytic strategy is far less complex and involves a lipid- and dynamin-dependent mechanism of entry. Collectively, the studies of PV and CVB3 entry into polarized versus non-polarized cells highlight the important differences that likely exist between these differing cell types.

Echoviruses

Studies of the entry of diverse echoviruses have been performed in both polarized and non-polarized cells. In non-polarized monkey kidney CV-1 cells, cholera toxin colocalized with echovirus 1 (EV1) and caveolin shortly after entry (Pietiäinen *et al.*, 2004). The virus is rapidly internalized following receptor binding, in a clathrin, actin, and microtubule independent process that requires cholesterol, caveolin and dynamin (Pietiäinen *et al.*, 2004). In polarized Caco-2 cells, EV1 localizes with VLA-2 in discrete foci on the cell surface, and is internalized and localized to the early endosomes in the perinuclear region by 20–40 min following entry. The virus remains in endosomes through 2 hpi, at which time it colocalizes with late endosomal and lysosomal markers (Krieger *et al.*, 2013). The entry process requires cholesterol, but caveolin depletion had no effect on internalization and instead entry was associated with macropinocytosis (Krieger *et al.*, 2013). However, dynamin II was also required for infection, and it is unclear what role dynamin plays in EV1 viral entry in the epithelium (Krieger *et al.*, 2013).

In the case of EV7, viral entry into polarized Caco-2 cells occurs prior to 1 hpi and requires clathrin (Kim and Bergelson, 2012). This is unlike CVB3 or EV1 entry into these same cells, which were both independent of CME. EV7 vRNA is released between 1 and 2 hpi, a time when the virus is transitioning from the early endosome in the perinuclear region to the lysosome (Kim and Bergelson, 2012, 2014). Rab7 is a protein that is involved in maturation of the endosome to the lysosome, and siRNA to or dominant negative forms of Rab7 had an inhibitory effect on EV7 infection, suggesting that vesicle maturation is required for infection. It is unclear why the virus would need to traffic to the late endosome or lysosome as the capsid is stable in acidic environments and cathepsin inhibition had no effect on infection (Kim and Bergelson, 2012). Rab7 is not only involved in endosomal maturation, but also in fusion between the autophagosome and lysosomes. 3-MA, an autophagy inhibitor that also inhibits recruitment of EEA1 and Rab5 to the early endosome, decreased infection, and if viruses were labelled with neutral red, a light pulse at 90 mpi decreased infection of treated, but not mock-treated cells. The 3-MA treated cells had accumulation of virus in large vesicles that were not labelled with the typical early endosome markers EEA-1 and Rab5 (Kim and Bergelson, 2014), suggesting that 3-MA inhibitory effect is on maturation of the endosome and not via autophagy. Consistent with this, while Caco-2 cells have a high level of autophagy, EV7 infection does not affect the basal rate of autophagy (Kim and Bergelson, 2014).

Concluding thoughts

Significant progress has been made regarding the identification of *Enterovirus* receptors and the characterization of the steps associated with entry and genome release. However, an important aspect of viral entry that must be considered when studying Enteroviruses is the fact that these viruses will enter the human host in the GI tract, which is a complex and microbiologically diverse environment. It is clear that this environment will impact a variety of aspects of *Enterovirus* entry and studies suggest that the bacterial microbiome present in the GI tract may be required to facilitate this process (Kuss *et al.*, 2011; Robinson *et al.*, 2014). In addition, the cellular complexity of the intestinal surface, which is composed of at least seven distinct cell types, is also likely to contribute to *Enterovirus* entry and subsequent infection. Recent work suggests that some Enteroviruses might infect the intestinal surface in a cell-type specific manner (Drummond *et al.*, 2017), although the mechanistic basis for this targeting remains unclear. Future studies focused on the specific trigger for genome release and the impact of the cellular and bacterial diversity present in the intestine will provide additional insights into the shared and unique pathways utilized by Enteroviruses to enter the human host.

References

Alexander, D.A., and Dimock, K. (2002). Sialic acid functions in enterovirus 70 binding and infection. J. Virol. *76*, 11265–11272.

Belnap, D.M., Filman, D.J., Trus, B.L., Cheng, N., Booy, F.P., Conway, J.F., Curry, S., Hiremath, C.N., Tsang, S.K., Steven, A.C., *et al.* (2000a). Molecular tectonic model of virus structural transitions: the putative cell entry states of poliovirus. J. Virol. *74*, 1342–1354.

Belnap, D.M., McDermott, B.M., Filman, D.J., Cheng, N., Trus, B.L., Zuccola, H.J., Racaniello, V.R., Hogle, J.M., and Steven, A.C. (2000b). Three-dimensional structure of poliovirus receptor bound to poliovirus. Proc. Natl. Acad. Sci. U.S.A. *97*, 73–78.

Bergelson, J.M., Chan, M., Solomon, K.R., St John, N.F., Lin, H. & Finberg, R.W. (1994). Decay-accelerating factor (CD55), a glycosylphosphatidylinositol-anchored complement regulatory protein, is a receptor for several echoviruses. *Proc Natl Acad Sci U S A*, 91, 6245–6248.

Bergelson, J.M., Cunningham, J.A., Droguett, G., Kurt-Jones, E.A., Krithivas, A., Hong, J.S., Horwitz, M.S., Crowell, R.L., and Finberg, R.W. (1997). Isolation of a common receptor for Coxsackie B viruses and adenoviruses 2 and 5. Science *275*, 1320–1323.

Bergelson, J.M., Mohanty, J.G., Crowell, R.L., St John, N.F., Lublin, D.M., and Finberg, R.W. (1995). Coxsackievirus B3 adapted to growth in RD cells binds to decay-accelerating factor (CD55). J. Virol. *69*, 1903–1906.

Bergelson, J.M., Shepley, M.P., Chan, B.M., Hemler, M.E., and Finberg, R.W. (1992). Identification of the integrin VLA-2 as a receptor for echovirus 1. Science *255*, 1718–1720.

Bochkov, Y.A., Watters, K., Ashraf, S., Griggs, T.F., Devries, M.K., Jackson, D.J., Palmenberg, A.C., and Gern, J.E. (2015). Cadherin-related family member 3, a childhood asthma susceptibility gene product, mediates rhinovirus C binding and replication. *Proc Natl Acad Sci U S A*, 112, 5485–5490.

Boonyakiat, Y., Hughes, P.J., Ghazi, F., and Stanway, G. (2001). Arginine-glycine-aspartic acid motif is critical for human parechovirus 1 entry. J. Virol. *75*, 10000–10004. https://doi.org/10.1128/JVI.75.20.10000-10004.2001

Bostina, M., Levy, H., Filman, D.J., and Hogle, J.M. (2011). Poliovirus RNA is released from the capsid near a twofold symmetry axis. J. Virol. *85*, 776–783. https://doi.org/10.1128/JVI.00531-10

Brandenburg, B., Lee, L.Y., Lakadamyali, M., Rust, M.J., Zhuang, X., and Hogle, J.M. (2007). Imaging poliovirus entry in live cells. PLOS Biol. *5*, e183.

Butan, C., Filman, D.J., and Hogle, J.M. (2014). Cryo-electron microscopy reconstruction shows poliovirus 135S particles poised for membrane interaction and RNA release. *J Virol*, 88, 1758–1770.

Chen, Z., Fischer, E.R., Kouiavskaia, D., Hansen, B.T., Ludtke, S.J., Bidzhieva, B., Makiya, M., Agulto, L., Purcell, R.H., and Chumakov, K. (2013). Cross-neutralizing human anti-poliovirus antibodies bind the recognition site for cellular receptor. Proc. Natl. Acad. Sci. U.S.A. *110*, 20242–20247. https://doi.org/10.1073/pnas.1320041110

Chow, M., Newman, J.F., Filman, D., Hogle, J.M., Rowlands, D.J., and Brown, F. (1987). Myristylation of picornavirus capsid protein VP4 and its structural significance. Nature *327*, 482–486. https://doi.org/10.1038/327482a0

Colonno, R.J., Condra, J.H., Mizutani, S., Callahan, P.L., Davies, M.E., and Murcko, M.A. (1988). Evidence for the direct involvement of the rhinovirus canyon in receptor binding. Proc. Natl. Acad. Sci. U.S.A. *85*, 5449–5453.

Coyne, C.B., and Bergelson, J.M. (2006). Virus-induced Abl and Fyn kinase signals permit coxsackievirus entry through epithelial tight junctions. *Cell*, 124, 119–131.

Coyne, C.B., Kim, K.S., and Bergelson, J.M. (2007). Poliovirus entry into human brain microvascular cells requires receptor-induced activation of SHP-2. *The EMBO journal*, 26, 4016–4028.

Crowell, R.L., and Philipson, L. (1971). Specific alterations of coxsackievirus B3 eluted from HeLa cells. J. Virol. *8*, 509–515.

Crowther, D., and Melnick, J.L. (1961). The incorporation of neutral red and acridine orange into developing poliovirus particles making them photosensitive. Virology *14*, 11–21.

Danthi, P., Tosteson, M., Li, Q.H., and Chow, M. (2003). Genome delivery and ion channel properties are altered in VP4 mutants of poliovirus. J. Virol. *77*, 5266–5274.

Drummond, C.G., Bolock, A.M., Ma, C., Luke, C.J., Good, M., and Coyne, C.B. (2017). Enteroviruses infect human enteroids and induce antiviral signaling in a cell lineage-specific manner. *Proc Natl Acad Sci U S A*, 114, 1672–1677.

Drummond, C.G., Nickerson, C.A., and Coyne, C.B. (2016). A Three-Dimensional Cell Culture Model To Study Enterovirus Infection of Polarized Intestinal Epithelial Cells. *mSphere*, 1.

Du, N., Cong, H., Tian, H., Zhang, H., Zhang, W., Song, L., and Tien, P. (2014). Cell surface vimentin is an attachment receptor for enterovirus 71. J. Virol. *88*, 5816–5833. https://doi.org/10.1128/JVI.03826-13

Elkin, S.R., Lakoduk, A.M., and Schmid, S.L. (2016). Endocytic pathways and endosomal trafficking: a primer. Wien Med Wochenschr *166*, 196–204. https://doi.org/10.1007/s10354-016-0432-7

Filman, D.J., Syed, R., Chow, M., Macadam, A.J., Minor, P.D., and Hogle, J.M. (1989). Structural factors that control conformational transitions and serotype specificity in type 3 poliovirus. EMBO J. *8*, 1567–1579.

Fricks, C.E., and Hogle, J.M. (1990). Cell-induced conformational change in poliovirus: externalization of the amino terminus of VP1 is responsible for liposome binding. *J. Virology*, 64, 1934–1945.

Garriga, D., Pickl-Herk, A., Luque, D., Wruss, J., Castón, J.R., Blaas, D., and Verdaguer, N. (2012). Insights into minor group rhinovirus uncoating: the X-ray structure of the HRV2 empty capsid. PLOS Pathog. *8*, e1002473. https://doi.org/10.1371/journal.ppat.1002473

Goodfellow, I.G., Sioofy, A.B., Powell, R.M., and Evans, D.J. (2001). Echoviruses bind heparan sulfate at the cell surface. J. Virol. *75*, 4918–4921. https://doi.org/10.1128/JVI.75.10.4918-4921.2001

Grant, R.A., Hiremath, C.N., Filman, D.J., Syed, R., Andries, K., and Hogle, J.M. (1994). Structures of poliovirus complexes with anti-viral drugs: implications for viral stability and drug design. Curr. Biol. *4*, 784–797.

Greve, J.M., Davis, G., Meyer, A.M., Forte, C.P., Yost, S.C., Marlor, C.W., Kamarck, M.E., and McClelland, A. (1989). The major human rhinovirus receptor is ICAM-1. Cell *56*, 839–847.

Groppelli, E., Levy, H.C., Sun, E., Strauss, M., Nicol, C., Gold, S., Zhuang, X., Tuthill, T.J., Hogle, J.M., and Rowlands, D.J. (2017). Picornavirus RNA is protected from cleavage by ribonuclease during virion uncoating and transfer across cellular and model membranes. PLOS Pathog. *13*, e1006197. https://doi.org/10.1371/journal.ppat.1006197

Hafenstein, S., Bowman, V.D., Chipman, P.R., Bator Kelly, C.M., Lin, F., Medof, M.E., and Rossmann, M.G. (2007). Interaction of decay-accelerating factor with coxsackievirus B3. J. Virol. *81*, 12927–12935.

Harutyunyan, S., Kowalski, H., and Blaas, D. (2014). The Rhinovirus subviral a-particle exposes 3'-terminal sequences of its genomic RNA. J. Virol. *88*, 6307–6317. https://doi.org/10.1128/JVI.00539-14

Harutyunyan, S., Kumar, M., Sedivy, A., Subirats, X., Kowalski, H., Köhler, G., and Blaas, D. (2013). Viral uncoating is directional: exit of the genomic RNA in a common cold virus starts with the poly-(A) tail at the 3'-end. PLOS Pathog. *9*, e1003270. https://doi.org/10.1371/journal.ppat.1003270

He, Y., Bowman, V.D., Mueller, S., Bator, C.M., Bella, J., Peng, X., Baker, T.S., Wimmer, E., Kuhn, R.J., and Rossmann, M.G. (2000). Interaction of the poliovirus receptor with poliovirus. Proc. Natl. Acad. Sci. U.S.A. *97*, 79–84.

He, Y., Chipman, P.R., Howitt, J., Bator, C.M., Whitt, M.A., Baker, T.S., Kuhn, R.J., Anderson, C.W., Freimuth, P., and Rossmann, M.G. (2001). Interaction of coxsackievirus B3 with the full length coxsackievirus-adenovirus receptor. Nat. Struct. Biol. *8*, 874–878. https://doi.org/10.1038/nsb1001-874

He, Y., Lin, F., Chipman, P.R., Bator, C.M., Baker, T.S., Shoham, M., Kuhn, R.J., Medof, M.E., and Rossmann, M.G. (2002). Structure of decay-accelerating factor bound to echovirus 7: a virus-receptor complex. Proc. Natl. Acad. Sci. U.S.A. *99*, 10325–10329. https://doi.org/10.1073/pnas.152161599

Heikkila, O., Susi, P., Stanway, G., and Hyypia, T. (2009). Integrin alphaVbeta6 is a high-affinity receptor for coxsackievirus A9. *J Gen Virol*, 90, 197–204.

Hewat, E.A., Neumann, E., Conway, J.F., Moser, R., Ronacher, B., Marlovits, T.C., and Blaas, D. (2000). The cellular receptor to human rhinovirus 2 binds around the 5-fold axis and not in the canyon: a structural view. EMBO J. *19*, 6317–6325. https://doi.org/10.1093/emboj/19.23.6317

Hofer, F., Gruenberger, M., Kowalski, H., Machat, H., Huettinger, M., Kuechler, E., and Blaas, D. (1994). Members of the low density lipoprotein receptor family mediate cell entry of a minor-group common cold virus. Proc. Natl. Acad. Sci. U.S.A. *91*, 1839–1842.

Hogle, J.M., Chow, M., and Filman, D.J. (1985). Three-dimensional structure of poliovirus at 2.9 A resolution. Science *229*, 1358–1365.

Holland, J.J. (1961). Receptor affinities as major determinants of enterovirus tissue tropisms in humans. Virology *15*, 312–326.

Holland, J.J., Mclaren, L.C., and Syverton, J.T. (1959). The mammalian cell-virus relationship. IV. Infection of naturally insusceptible cells with enterovirus ribonucleic acid. *The Journal of Experimental Medicine*, 110, 65–80.

Hsu, K.H., Lonberg-Holm, K., Alstein, B., and Crowell, R.L. (1988). A monoclonal antibody specific for the cellular receptor for the group B coxsackieviruses. J. Virol. *62*, 1647–1652.

Imamura, T., Okamoto, M., Nakakita, S., Suzuki, A., Saito, M., Tamaki, R., Lupisan, S., Roy, C.N., Hiramatsu, H., Sugawara, K.E., *et al.* (2014). Antigenic and receptor binding properties of enterovirus 68. J. Virol. *88*, 2374–2384. https://doi.org/10.1128/JVI.03070-13

Karnauchow, T.M., Tolson, D.L., Harrison, B.A., Altman, E., Lublin, D.M., and Dimock, K. (1996). The HeLa cell receptor for enterovirus 70 is decay-accelerating factor (CD55). J. Virol. *70*, 5143–5152.

Kim, C., and Bergelson, J.M. (2012). Echovirus 7 entry into polarized intestinal epithelial cells requires clathrin and Rab7. MBio *3*, e00304–11. https://doi.org/10.1128/mBio.00304-11

Kim, C., and Bergelson, J.M. (2014). Echovirus 7 entry into polarized caco-2 intestinal epithelial cells involves core components of the autophagy machinery. J. Virol. *88*, 434–443. https://doi.org/10.1128/JVI.02706-13

King, S.L., Cunningham, J.A., Finberg, R.W., and Bergelson, J.M. (1995). Echovirus 1 interaction with the isolated VLA-2 I domain. J. Virol. *69*, 3237–3239.

Kolatkar, P.R., Bella, J., Olson, N.H., Bator, C.M., Baker, T.S., and Rossmann, M.G. (1999). Structural studies of two rhinovirus serotypes complexed with fragments of their cellular receptor. EMBO J. *18*, 6249–6259. https://doi.org/10.1093/emboj/18.22.6249

Korant, B.D., Lonberg-Holm, K., Noble, J., and Stasny, J.T. (1972). Naturally occurring and artificially produced components of three rhinoviruses. Virology *48*, 71–86.

Krieger, S.E., Kim, C., Zhang, L., Marjomaki, V., and Bergelson, J.M. (2013). Echovirus 1 entry into polarized Caco-2 cells depends on dynamin, cholesterol, and cellular factors associated with macropinocytosis. J. Virol. *87*, 8884–8895. https://doi.org/10.1128/JVI.03415-12

Kuss, S.K., Best, G.T., Etheredge, C.A., Pruijssers, A.J., Frierson, J.M., Hooper, L.V., Dermody, T.S., and Pfeiffer, J.K. (2011). Intestinal microbiota promote enteric virus replication and systemic pathogenesis. *Science*, 334, 249–252.

Lee, H., Shingler, K.L., Organtini, L.J., Ashley, R.E., Makhov, A.M., Conway, J.F., and Hafenstein, S. (2016). The novel asymmetric entry intermediate of a picornavirus captured with nanodiscs. Sci Adv *2*, e1501929. https://doi.org/10.1126/sciadv.1501929

Liu, Y., Sheng, J., Baggen, J., Meng, G., Xiao, C., Thibaut, H.J., van Kuppeveld, F.J., and Rossmann, M.G. (2015). Sialic acid-dependent cell entry of human enterovirus D68. Nat Commun *6*, 8865. https://doi.org/10.1038/ncomms9865

Madshus, I.H., Olsnes, S., and Sandvig, K. (1984). Mechanism of entry into the cytosol of poliovirus type 1: requirement for low pH. J. Cell Biol. *98*, 1194–1200.

Maxfield, F.R., and McGraw, T.E. (2004). Endocytic recycling. Nat. Rev. Mol. Cell Biol. *5*, 121–132. https://doi.org/10.1038/nrm1315

Mayor, S., Parton, R.G., and Donaldson, J.G. (2014). Clathrin-independent pathways of endocytosis. Cold Spring Harb. Perspect. Biol. *6*, a016758. https://doi.org/10.1101/cshperspect.a016758

McLaren, L.C., Holland, J.J., and Syverton, J.T. (1959). The mammalian cell-virus relationship. I. Attachment of poliovirus to cultivated cells of primate and non-primate origin. J. Exp. Med. *109*, 475–485.

McLeish, N.J., Williams, Ç.H., Kaloudas, D., Roivainen, M.M., and Stanway, G. (2012). Symmetry-related clustering of positive charges is a common mechanism for heparan sulfate binding in enteroviruses. J. Virol. *86*, 11163–11170.

Mendelsohn, C.L., Wimmer, E., and Racaniello, V.R. (1989). Cellular receptor for poliovirus: molecular cloning, nucleotide sequence, and expression of a new member of the immunoglobulin superfamily. *Cell*, *56*, 855–865.

Merilahti, P., Karelehto, E., and Susi, P. (2016). Role of Heparan Sulfate in Cellular Infection of Integrin-Binding Coxsackievirus A9 and Human Parechovirus 1 Isolates. PLOS ONE *11*, e0147168. https://doi.org/10.1371/journal.pone.0147168

Milstone, A., Petrella, J., Sanchez, M., Mahmud, M., Whitbeck, J., and Bergelson, J. (2005). Interaction with coxsackievirus and adenovirus receptor (CAR), but not with decay accelerating factor (DAF), induces A particle formation in a DAF-binding coxsackievirus B3 isolate. *J. Virol.*, *79*, 655–660.

Moscufo, N., Yafal, A.G., Rogove, A., Hogle, J., and Chow, M. (1993). A mutation in VP4 defines a new step in the late stages of cell entry by poliovirus. J. Virol. *67*, 5075–5078.

Muckelbauer, J.K., Kremer, M., Minor, I., Diana, G., Dutko, F.J., Groarke, J., Pevear, D.C., and Rossmann, M.G. (1995). The structure of coxsackievirus B3 at 3.5 A resolution. Structure *3*, 653–667.

Müller, S., Dennemärker, J., and Reinheckel, T. (2012). Specific functions of lysosomal proteases in endocytic and autophagic pathways. Biochim. Biophys. Acta *1824*, 34–43. https://doi.org/10.1016/j.bbapap.2011.07.003

Nilsson, E.C., Jamshidi, F., Johansson, S.M., Oberste, M.S., and Arnberg, N. (2008). Sialic acid is a cellular receptor for coxsackievirus A24 variant, an emerging virus with pandemic potential. J. Virol. *82*, 3061–3068. https://doi.org/10.1128/JVI.02470-07

Nishimura, Y., Lee, H., Hafenstein, S., Kataoka, C., Wakita, T., Bergelson, J.M., and Shimizu, H. (2013). Enterovirus 71 binding to PSGL-1 on leukocytes: VP1-145 acts as a molecular switch to control receptor interaction. PLOS Pathog. *9*, e1003511. https://doi.org/10.1371/journal.ppat.1003511

Nishimura, Y., McLaughlin, N.P., Pan, J., Goldstein, S., Hafenstein, S., Shimizu, H., Winkler, J.D., and Bergelson, J.M. (2015). The Suramin Derivative NF449 Interacts with the 5-fold Vertex of the Enterovirus A71 Capsid to Prevent Virus Attachment to PSGL-1 and Heparan Sulfate. PLOS Pathog. *11*, e1005184. https://doi.org/10.1371/journal.ppat.1005184

Nishimura, Y., Shimojima, M., Tano, Y., Miyamura, T., Wakita, T., and Shimizu, H. (2009). Human P-selectin glycoprotein ligand-1 is a functional receptor for enterovirus 71. Nat. Med. *15*, 794–797. https://doi.org/10.1038/nm.1961

Numrich, J., and Ungermann, C. (2014). Endocytic Rabs in membrane trafficking and signaling. Biol. Chem. *395*, 327–333. https://doi.org/10.1515/hsz-2013-0258

Nurani, G., Lindqvist, B., and Casasnovas, J.M. (2003). Receptor priming of major group human rhinoviruses for uncoating and entry at mild low-pH environments. J. Virol. *77*, 11985–11991.

Olson, N.H., Kolatkar, P.R., Oliveira, M.A., Cheng, R.H., Greve, J.M., McClelland, A., Baker, T.S., and Rossmann, M.G. (1993). Structure of a human rhinovirus complexed with its receptor molecule. Proc. Natl. Acad. Sci. U.S.A. *90*, 507–511.

Patel, K.P., Coyne, C.B., and Bergelson, J.M. (2009). Dynamin- and lipid raft-dependent entry of decay-accelerating factor (DAF)-binding and non-DAF-binding coxsackieviruses into nonpolarized cells. J. Virol. *83*, 11064–11077. https://doi.org/10.1128/JVI.01016-09

Pérez, L., and Carrasco, L. (1993). Entry of poliovirus into cells does not require a low-pH step. J. Virol. *67*, 4543–4548.

Pietiäinen, V., Marjomäki, V., Upla, P., Pelkmans, L., Helenius, A., and Hyypiä, T. (2004). Echovirus 1 endocytosis into caveosomes requires lipid rafts, dynamin II, and signaling events. Mol. Biol. Cell *15*, 4911–4925. https://doi.org/10.1091/mbc.E04-01-0070

Powell, R.M., Schmitt, V., Ward, T., Goodfellow, I., Evans, D.J., and Almond, J.W. (1998). Characterization of echoviruses that bind decay accelerating factor (CD55): evidence that some haemagglutinating strains use more than one cellular receptor. J. Gen. Virol. *79*, 1707–1713. https://doi.org/10.1099/0022-1317-79-7-1707

Powell, R.M., Ward, T., Evans, D.J., and Almond, J.W. (1997). Interaction between echovirus 7 and its receptor, decay-accelerating factor (CD55): evidence for a secondary cellular factor in A-particle formation. *J. Virology*, 71, 9306–9312.

Powell, R.M., Ward, T., Goodfellow, I., Almond, J.W., and Evans, D.J. (1999. Mapping the binding domains on decay accelerating factor (DAF) for haemagglutinating enteroviruses: implications for the evolution of a DAF-binding phenotype. *J Gen Virol*, 80 (Pt 12), 3145–3152.

Ren, J., Wang, X., Hu, Z., Gao, Q., Sun, Y., Li, X., Porta, C., Walter, T.S., Gilbert, R.J., Zhao, Y., *et al.* (2013). Picornavirus uncoating intermediate captured in atomic detail. Nat Commun *4*, 1929. https://doi.org/10.1038/ncomms2889

Robinson, C.M., Jesudhasan, P.R., and Pfeiffer, J.K. (2014). Bacterial lipopolysaccharide binding enhances virion stability and promotes environmental fitness of an enteric virus. *Cell Host Microbe*, 15, 36-46.

Rogers, J.M., Diana, G.D., and McKinlay, M.A. (1999). Pleconaril. A broad spectrum antipicornaviral agent. Adv. Exp. Med. Biol. *458*, 69–76.

Roivainen, M., Hyypiä, T., Piirainen, L., Kalkkinen, N., Stanway, G., and Hovi, T. (1991). RGD-dependent entry of coxsackievirus A9 into host cells and its bypass after cleavage of VP1 protein by intestinal proteases. J. Virol. *65*, 4735–4740.

Roivainen, M., Piirainen, L., Hovi, T., Virtanen, I., Riikonen, T., Heino, J., and Hyypiä, T. (1994). Entry of coxsackievirus A9 into host cells: specific interactions with alpha v beta 3 integrin, the vitronectin receptor. Virology *203*, 357–365.

Rossmann, M.G. (1989). The canyon hypothesis. Hiding the host cell receptor attachment site on a viral surface from immune surveillance. J. Biol. Chem. *264*, 14587–14590.

Rossmann, M.G., Arnold, E., Erickson, J.W., Frankenberger, E.A., Griffith, J.P., Hecht, H.-J., Johnson, J.E., Kamer, G., Luo, M., Mosser, A.G., Rueckert, R.R., Sherry, B., and Vriend, G. (1985). Structure of a human common cold virus and functional relationship to other picornaviruses. *Nature*, 317, 145–153.

Seitsonen, J., Susi, P., Heikkilä, O., Sinkovits, R.S., Laurinmäki, P., Hyypiä, T., and Butcher, S.J. (2010). Interaction of alphaVbeta3 and alphaVbeta6 integrins with human parechovirus 1. J. Virol. *84*, 8509–8519. https://doi.org/10.1128/JVI.02176-09

Shafren, D.R., Bates, R.C., Agrez, M.V., Herd, R.L., Burns, G.F., and Barry, R.D. (1995). Coxsackieviruses B1, B3, and B5 use decay accelerating factor as a receptor for cell attachment. J. Virol. *69*, 3873–3877.

Shafren, D.R., Dorahy, D.J., Greive, S.J., Burns, G.F., and Barry, R.D. (1997a). Mouse cells expressing human intercellular adhesion molecule-1 are susceptible to infection by coxsackievirus A21. J. Virol. *71*, 785–789.

Shafren, D.R., Dorahy, D.J., Ingham, R.A., Burns, G.F., and Barry, R.D. (1997b). Coxsackievirus A21 binds to decay-accelerating factor but requires intercellular adhesion molecule 1 for cell entry. *J Virol*, 71, 4736–4743.

Shieh, J.T., and Bergelson, J.M. (2002. Interaction with decay-accelerating factor facilitates coxsackievirus B infection of polarized epithelial cells. *J Virol*, 76, 9474–9480.

Smith, T.J., Chase, E.S., Schmidt, T.J., Olson, N.H., and Baker, T.S. (1996). Neutralizing antibody to human rhinovirus 14 penetrates the receptor-binding canyon. Nature *383*, 350–354. https://doi.org/10.1038/383350a0

Staring, J., von Castelmur, E., Blomen, V.A., van den Hengel, L.G., Brockmann, M., Baggen, J., Thibaut, H.J., Nieuwenhuis, J., Janssen, H., van Kuppeveld, F.J., *et al.* (2017). PLA2G16 represents a switch between entry and clearance of Picornaviridae. Nature *541*, 412–416. https://doi.org/10.1038/nature21032

Staunton, D.E., Merluzzi, V.J., Rothlein, R., Barton, R., Marlin, S.D., and Springer, T.A. (1989). A cell adhesion molecule, ICAM-1, is the major surface receptor for rhinoviruses. Cell *56*, 849–853.

Stins, M.F., Badger, J., and Sik Kim, K. (2001). Bacterial invasion and transcytosis in transfected human brain microvascular endothelial cells. *Microb Pathog*, 30, 19–28.

Su, P.Y., Wang, Y.F., Huang, S.W., Lo, Y.C., Wang, Y.H., Wu, S.R., Shieh, D.B., Chen, S.H., Wang, J.R., Lai, M.D., *et al.* (2015). Cell surface nucleolin facilitates enterovirus 71 binding and infection. J. Virol. *89*, 4527–4538. https://doi.org/10.1128/JVI.03498-14

Tan, C.W., Poh, C.L., Sam, I.C., and Chan, Y.F. (2013). Enterovirus 71 uses cell surface heparan sulfate glycosaminoglycan as an attachment receptor. J. Virol. *87*, 611–620. https://doi.org/10.1128/JVI.02226-12

Tomassini, J.E., Graham, D., DeWitt, C.M., Lineberger, D.W., Rodkey, J.A., and Colonno, R.J. (1989). cDNA cloning reveals that the major group rhinovirus receptor on HeLa cells is intercellular adhesion molecule 1. Proc. Natl. Acad. Sci. U.S.A. *86*, 4907–4911.

Tomko, R.P., Xu, R., and Philipson, L. (1997). HCAR and MCAR: the human and mouse cellular receptors for subgroup C adenoviruses and group B coxsackieviruses. Proc. Natl. Acad. Sci. U.S.A. *94*, 3352–3356.

Tosteson, M.T., and Chow, M. (1997). Characterization of the ion channels formed by poliovirus in planar lipid membranes. J. Virol. *71*, 507–511.

Triantafilou, K., Fradelizi, D., Wilson, K., and Triantafilou, M. (2002). GRP78, a coreceptor for coxsackievirus A9, interacts with major histocompatibility complex class I molecules which mediate virus internalization. *J Virol*, 76, 633–643.

Triantafilou, K., Triantafilou, M., Takada, Y., and Fernandez, N. (2000). Human parechovirus 1 utilizes integrins alphavbeta3 and alphavbeta1 as receptors. J. Virol. *74*, 5856–5862.
Uncapher, C.R., DeWitt, C.M., and Colonno, R.J. (1991). The major and minor group receptor families contain all but one human rhinovirus serotype. Virology *180*, 814–817.
Wandinger-Ness, A., and Zerial, M. (2014). Rab proteins and the compartmentalization of the endosomal system. Cold Spring Harb Perspect Biol 6, a022616. https://doi.org/10.1101/cshperspect.a022616
Ward, T., Pipkin, P.A., Clarkson, N.A., Stone, D.M., Minor, P.D., and Almond, J.W. (1994). Decay-accelerating factor CD55 is identified as the receptor for echovirus 7 using CELICS, a rapid immuno-focal cloning method. EMBO J. *13*, 5070–5074.
Wei, W., Guo, H., Chang, J., Yu, Y., Liu, G., Zhang, N., Willard, S.H., Zheng, S., and Yu, X.F. (2016). ICAM-5/Telencephalin Is a Functional Entry Receptor for Enterovirus D68. Cell Host Microbe *20*, 631–641.
Xing, L., Casasnovas, J.M., and Cheng, R.H. (2003). Structural analysis of human rhinovirus complexed with ICAM-1 reveals the dynamics of receptor-mediated virus uncoating. J. Virol. *77*, 6101–6107.
Xing, L., Tjarnlund, K., Lindqvist, B., Kaplan, G.G., Feigelstock, D., Cheng, R.H., and Casasnovas, J.M. (2000). Distinct cellular receptor interactions in poliovirus and rhinoviruses. EMBO J. *19*, 1207–1216. https://doi.org/10.1093/emboj/19.6.1207
Yafal, A.G., Kaplan, G., Racaniello, V.R., and Hogle, J.M. (1993). Characterization of poliovirus conformational alteration mediated by soluble cell receptors. Virology *197*, 501–505.
Yamayoshi, S., Iizuka, S., Yamashita, T., Minagawa, H., Mizuta, K., Okamoto, M., Nishimura, H., Sanjoh, K., Katsushima, N., Itagaki, T., *et al.* (2012). Human SCARB2-dependent infection by coxsackievirus A7, A14, and A16 and enterovirus 71. J. Virol. *86*, 5686–5696. https://doi.org/10.1128/JVI.00020-12
Yamayoshi, S., Ohka, S., Fujii, K., and Koike, S. (2013). Functional comparison of SCARB2 and PSGL1 as receptors for enterovirus 71. J. Virol. *87*, 3335–3347. https://doi.org/10.1128/JVI.02070-12
Yamayoshi, S., Yamashita, Y., Li, J., Hanagata, N., Minowa, T., Takemura, T., and Koike, S. (2009). Scavenger receptor B2 is a cellular receptor for enterovirus 71. Nat. Med. *15*, 798–801. https://doi.org/10.1038/nm.1992
Yang, S.L., Chou, Y.T., Wu, C.N., and Ho, M.S. (2011). Annexin II binds to capsid protein VP1 of enterovirus 71 and enhances viral infectivity. J. Virol. *85*, 11809–11820. https://doi.org/10.1128/JVI.00297-11
Ylipaasto, P., Eskelinen, M., Salmela, K., Hovi, T., and Roivainen, M. (2010). Vitronectin receptors, alpha v integrins, are recognized by several non-RGD-containing echoviruses in a continuous laboratory cell line and also in primary human Langerhans' islets and endothelial cells. *J Gen Virol*, 91, 155–165.
Zautner, A.E., Körner, U., Henke, A., Badorff, C., and Schmidtke, M. (2003). Heparan sulfates and coxsackievirus-adenovirus receptor: each one mediates coxsackievirus B3 PD infection. J. Virol. *77*, 10071–10077.

Hijacking Host Functions for Translation and RNA Replication by Enteroviruses

3

Sonia Maciejewski and Bert L. Semler*

Department of Microbiology and Molecular Genetics, School of Medicine, University of California, Irvine, CA, USA.

*Correspondence: blsemler@uci.edu

https://doi.org/10.21775/9781910190739.03

Abstract

The *Enterovirus* genus includes the species poliovirus, coxsackievirus, rhinovirus, and enterovirus. These viruses can cause severe diseases in certain individuals, including poliomyelitis, myocarditis, and meningitis. Rhinovirus is responsible for one of the most prevalent human diseases in the world, the common cold. Although diseases caused by these infections can be severe, no antiviral against Enteroviruses is currently available. To develop broad-spectrum antivirals, the molecular components and mechanistic steps of the viral replication cycle must be identified. Due to the small genomic RNA ($\approx$7.5 kb) of Enteroviruses, host proteins are utilized to mediate viral replication. Although some of these cellular proteins have been identified and their roles in picornavirus replication have been characterized, it is necessary to identify and elucidate the replication functions of additional cellular proteins to develop new potential targets for antiviral therapeutics. Enteroviruses are known to modify cellular proteins to stimulate their levels of gene expression and RNA synthesis, but there are some cases where unaltered host proteins can aid in viral replication. Enteroviruses can also evade the antiviral response by altering host proteins involved in the immune and stress response to ensure efficient viral replication. How Enteroviruses modify and utilize these host proteins will be discussed in this chapter.

Introduction

The diverse *Enterovirus* genus of the *Picornaviridae* family encompasses 12 species, including poliovirus, coxsackievirus, rhinovirus, enterovirus, and echovirus serotypes (ictvonline.org/virusTaxonomy.asp). These viruses are responsible for the most prevalent human diseases worldwide (Khetsuriani *et al.*, 2006), such as the common cold. Other illnesses include poliomyelitis, pericarditis, myocarditis, hand, foot, and mouth disease, and meningitis, which can severely affect infants, the elderly, and immunocompromised individuals.

Although the symptoms of respiratory illnesses caused by picornaviruses are almost never fatal, these viral infections have a negative economic impact due to lost work time and can severely affect individuals with respiratory dysfunction, such as asthma (Gavala *et al.*, 2011). Such respiratory infections are commonly caused by human rhinovirus, coxsackievirus, and enterovirus D68. Enterovirus D68 was first identified in California in 1962, but has recently had recurring outbreaks in North America, Europe, and Asia (Tokarz *et al.*, 2012). Enterovirus D68 outbreaks have been associated with severe respiratory illnesses and are quickly spreading throughout the United States (Midgley *et al.*, 2014). Another *Enterovirus* with recurring outbreaks is enterovirus 71. Enterovirus 71 is a neurotropic *Enterovirus* with symptoms similar to hand, foot, and mouth disease and remains endemic in the Asia-Pacific region. Neurological diseases caused by enterovirus 71 infection can cause aseptic meningitis and brainstem encephalitis, which can lead to mortality (reviewed in Shih *et al.*, 2011). Although a vaccine against poliovirus is available, no effective antivirals for treating *Enterovirus* infections currently exist. Since symptoms caused by such infections can lead to severe complications in certain individuals, it is necessary to develop antiviral therapeutics against Enteroviruses. Antivirals can target a host protein required for *Enterovirus* replication, a viral protein, or the viral RNAs. Antivirals targeting the host can lead to cell toxicity, while antivirals against a viral protein can lead to antiviral-resistant mutants. To develop an effective broad-spectrum antiviral, the steps of the *Enterovirus* replication mechanism and the roles of key molecular players, both host and viral, must be elucidated.

While infection may result in diverse diseases, all Enteroviruses have a small (≈7.5 kb) positive-sense, single stranded RNA genome that is replicated in the cytoplasm of infected cells. The genome contains a highly structured 5′ non-coding region (NCR) that is necessary for viral translation and RNA replication. Following the 5′ NCR is the coding region, which encodes both the structural and non-structural proteins, a 3′ NCR, and a short genetically encoded poly(A) tract at the 3′ terminus (Kitamura *et al.*, 1981; Wimmer *et al.*, 1993; Yogo and Wimmer, 1972). Enteroviruses lack a 7-methylguanosine (7mG) cap at the 5′ end of their RNA and instead contain a small viral protein known as VPg covalently linked to the 5′ end by a tyrosyl-RNA phosphodiester bond (Ambros and Baltimore, 1978; Flanegan *et al.*, 1977; Lee *et al.*, 1977; Rothberg *et al.*, 1978). Enteroviruses have evolved to use VPg as a protein-primer for RNA synthesis, since their RNA-dependent RNA polymerase (RdRP) 3D ($3D^{pol}$) cannot initiate viral RNA replication *de novo* (Flanegan and Baltimore, 1977; Paul *et al.*, 1998). The position of VPg on the genomic RNA is outlined in Fig. 3.1, which depicts an overview of the *Enterovirus* genome and the structural and non-structural viral proteins it encodes.

Following virion uncoating after the onset of infection, *Enterovirus* genomic RNA is localized in the cell cytoplasm where it is then translated into a single viral polyprotein. Since *Enterovirus* genomic RNAs lack a 7mG cap at the 5′ end, the viral polyprotein is translated in a cap-independent manner via an internal ribosome entry site (IRES). The IRES is located in the 5′ NCR of viral RNA and is composed of a number of stem–loop secondary structures. Due to their limited coding capacity (≈7.5 kb), Enteroviruses have evolved to hijack host cellular functions to carry out the translation and replication of their genomes. Viral translation is mediated by cellular IRES trans-acting factors (ITAFs). After translation of the open reading frame of genomic RNA, the viral polyprotein is proteolytically processed by viral proteinases. These viral proteinases can also cleave host proteins, resulting in the shut down

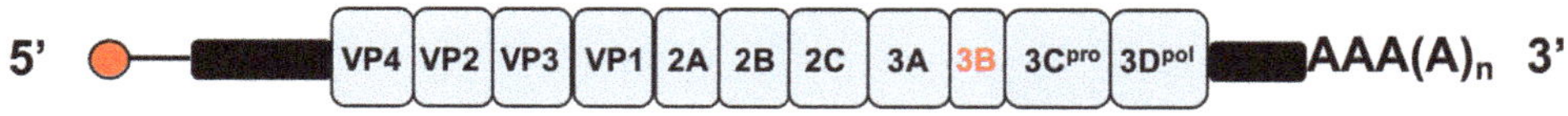

Figure 3.1 Overview of the *Enterovirus* genome. Enteroviruses have a small (≈7.5 kb), positive-sense RNA genome. At the 5′ end is the viral protein genome-linked (VPg) in red. VPg is covalently linked to the 5′ end of the RNA via an O4-(5′-uridylyl)tyrosine phosphodiester bond. VPg serves as a protein-primer for RNA synthesis since the RNA dependent RNA polymerase $3D^{pol}$ cannot initiate RNA synthesis *de novo*. The highly structured 5′ noncoding region (NCR) consists of stem–loops I–VI. Stem–loop I, also known as the cloverleaf, is required for viral RNA synthesis. Stem–loops II–VI encode the internal ribosome entry site (IRES), which initiates translation of the viral genome via a cap-independent mechanism. Following the 5′ NCR is the coding region. The coding region is translated into a single polyprotein, containing the structural (VP4-VP1) and non-structural (2A, 2B, 2C, 3A, 3B, 3C, and 3D) proteins. The polyprotein is then proteolytically processed by virus-encoded proteinases. The non-structural proteins are involved in multiple steps throughout the replication cycle, including modifying the host cell environment to aid in replication. Following the coding region is the 3′ NCR and genetically encoded poly(A) tract, which are necessary for efficient viral replication (reviewed in Bedard and Semler, 2004).

of cellular translation and transcription and the alteration of nucleo-cytoplasmic trafficking. These cleavage events are advantageous to the virus since the predominantly nuclear host proteins involved in viral replication become more concentrated in the cell cytoplasm. Following viral protein synthesis, specific viral proteins alter cytoplasmic membranes to form replication complexes, where viral RNA is synthesized. These newly synthesized RNAs can either undergo further rounds of translation and replication or become encapsidated into virions that go on to infect neighbouring cells. In addition to the modifications of host proteins for viral replication, viral proteinases can also alter cellular proteins to suppress antiviral response pathways, such as the type I interferon (IFN) response, generation of stress granules (SGs), and processing body (P body) formation. This review will focus on how Enteroviruses subvert host cellular proteins to enhance viral replication while evading antiviral and stress response pathways.

Viral proteinase disruption of host machinery

Enteroviruses can disrupt host cell translation and transcription machinery to benefit viral replication by using virus-encoded proteinases to cleave host cell proteins. To carry out these modifications, Enteroviruses utilize the virus-encoded proteinases 2A and 3C (and the precursor protein, 3CD). In addition to recognizing multiple cleavage sites in host cellular proteins, these proteins are responsible for proteolytically processing the *Enterovirus* polyprotein. *Poliovirus* proteinase 2A cleaves between phenylalanine-glycine or tyrosine-glycine residues in the viral polyprotein (Toyoda *et al.*, 1986), while 3C/3CD cleaves primarily at glutamine-glycine sites but can cleave at additional sites as well. 3C/3CD cleavage activity is dependent on surrounding sequences, specifically an amino acid with an aliphatic side chain in the amino acid located four positions (P4) proximal to the cleavage site (Blair and Semler, 1991; Nicklin *et al.*, 1986). The somewhat divergent recognition sites for these proteinases allow for cleavage of host proteins, which disrupts cellular functions and alters protein activities. This section will focus on how *Enterovirus* proteinases cleave host proteins to shut down the cellular translation and transcription machinery, subverting host functions to augment viral translation and RNA synthesis. These cleavage events and cellular function alterations are outlined in Table 3.1.

During cap-dependent translation of cellular mRNAs, eukaryotic initiation factors are recruited to the 7mG cap structure at the 5′ ends of mRNA. These factors form a complex that interacts with the 43S pre-initiation complex (PIC) to recruit ribosomes for translation initiation. Eukaryotic initiation factor 4G (eIF4G) serves as a scaffold protein that aids in the recruitment of eIF4E and eIF4A, to form a ribonucleoprotein (RNP) complex termed eIF4F, as well as additional proteins such as poly(A) binding protein (PABP), for initiation of cap-dependent translation (Jackson *et al.*, 2005; Wells *et al.*, 1998). Poliovirus and coxsackievirus 2A proteinases have been shown to cleave both eIF4G isoforms, eIF4GI and eIF4GII, early in viral infection, resulting in the loss of the N-terminal domain required for both eIF4E and PABP interaction [reviewed in Daijogo and Semler (2011)] (Devaney *et al.*, 1988; Etchison *et al.*, 1982; Kräusslich *et al.*, 1987). 2A proteinase preferentially cleaves eIF4G when the cellular protein is bound to cap-binding protein eIF4E, thus leading to rapid shut down of cap-dependent cellular translation (Bovee *et al.*, 1998). This host machinery shut down allows for resource allocation to efficient cap-independent translation benefiting viral protein synthesis. Additionally, evidence suggests that the cleaved form of eIF4G is required to stimulate IRES-dependent translation (Lamphear *et al.*, 1995; Liebig *et al.*, 1993). eIF4G interacts with stem–loop V of the poliovirus and coxsackievirus IRES (de Breyne *et al.*, 2009). The central domain of eIF4G interacts with eIF3, a component of the 43S PIC, *in vitro* (Sweeney *et al.*, 2014). This direct interaction between eIF4G and eIF3 at stem–loop V suggests that this interaction may be required in recruiting the 43S PIC to the proximal stem–loop IV, an essential step in 48S complex formation for viral translation initiation to occur.

Table 3.1 *Enterovirus* cleavage targets to disrupt the host cell translation and transcription machinery. Virus-encoded proteinases mediate cleavage of cellular proteins to shut down host cell functions, including cap-dependent translation and cellular transcription. Viral disruption allows host functions to become available for viral translation and RNA synthesis activities. The cellular proteins involved in cellular translation and transcription targeted by viral proteinases are outlined in this table

Host protein	Viral proteinase	Host functions disrupted	References
eIF4GI/II	PV and CVB3 2A	Cap-dependent translation	Devaney *et al.* (1988), Etchison *et al.* (1982), Kräusslich *et al.* (1987), Bovee *et al.* (1998), Lamphear *et al.* (1995)
PABP	PV and CVB3 2A, PV 3C	Cap-dependent translation	Joachims *et al.* (1999), Kuyumcu *et al.* (2002), Kerekatte *et al.* (1999)
UBF	PV 3C	RNA pol I transcription	Banerjee *et al.* (2005)
TAF_{110}	PV 3C	RNA pol I transcription	Banerjee *et al.* (2005)
TBP	PV 2A, PV 3C	PIC formation for RNA pol II transcription	Yalamanchili *et al.* (1997a), Das and Dasgupta (1993), Yalamanchili *et al.* (1996)
CREB-P	PV 3C	RNA pol II transcription	Yalamanchili *et al.* (1997b)
p53	PV 3C	Transcription	Weidman *et al.* (2001)
TFIIIC	PV 3C	RNA pol II transcription	Clark *et al.* (1991), Shen *et al.* (1996)

In addition to viral-mediated cleavage of the cap-binding complex scaffold protein eIF4G, *Enterovirus* infection also results in the cleavage of host protein PABP to disrupt cap-dependent translation. PABP is a cellular protein that binds to the 3′ poly(A) tract of mRNAs and interacts with eIF4G to functionally circularize the mRNA for efficient translation and mRNA stability in the uninfected cell [reviewed in (Fitzgerald and Semler, 2009; Smith *et al.*, 2014)]. During infection, PABP is cleaved by poliovirus and coxsackievirus 2A proteinases (Joachims *et al.*, 1999; Kerekatte *et al.*, 1999). In addition, poliovirus and human rhinovirus 3C proteinases preferentially cleave ribosome-associated PABP (Kuyumcu-Martinez *et al.*, 2002). PABP has three conserved putative cleavage sites in the flexible linker domain between its RNA recognition motifs (RRMs) and C-terminal domain and two additional putative cleavage sites in the RRMs (Kozlov *et al.*, 2001, 2004; Lloyd, 2006). Cleavage of PABP at these different sites by either 2A or 3C results in cellular translation inhibition by disrupting mRNA circularization.

Enterovirus proteinases 2A and 3C play roles in shutting down host cellular transcription during infection by disrupting RNA polymerases (pol) I, II, and III. *Poliovirus* proteinase 3C is responsible for inhibiting RNA pol I transcription activity approximately 90 to 180 minutes post infection by targeting the pol I transcription factor upstream binding factor (UBF), which is a sequence-specific DNA-binding protein that stabilizes the selectivity factor (SL-1) protein complex on the rRNA promoter for pol I transcription, and the SL-1 protein complex subunit, TATA-binding protein (TBP)- associated factor (TAF), TAF_{110} (Banerjee *et al.*, 2005). RNA pol II is responsible for transcribing host cellular mRNAs and is targeted by both *poliovirus* proteinases 2A and 3C. These enzymes cleave TBP, which is involved in forming a PIC containing transcription factor II D (TFIID) for pol II binding to transcription start sites (Das and Dasgupta, 1993; Yalamanchili *et al.*, 1997a). However, it is 3C-mediated cleavage of TBP and phosphorylated CREB, both upstream cellular transcription factors, that is required for pol II transcription inhibition (Das and Dasgupta, 1993; Yalamanchili *et al.*, 1996, 1997b). Poliovirus 3C activity can also lead to the degradation of transcription activator p53, in a non-ubiquitin mediated pathway, but in the presence of an unknown cellular protein (Weidman *et al.*, 2001). RNA pol III is responsible for transcribing ribosomal RNA genes, tRNA genes, and genes encoding other small RNAs. Pol III activity is inhibited during poliovirus infection by 3C-mediated cleavage of TFIIIC (Clark *et al.*, 1991; Shen *et al.*, 2004). TFIIIC binds to the promoter element, B box, downstream of the transcription start site to recruit TFIIIB, which recruits pol III to the transcription start site. Once TFIIIB recruits pol III, TFIIIC dissociates, allowing pol III-mediated transcription to occur. 3C cleavage of TFIIIC inhibits recruitment of TFIIIB, thus indirectly inhibiting pol III transcription (Clark *et al.*, 1991; Kassavetis *et al.*, 1990). These proteinase-mediated cleavages of host proteins all work together to inhibit cellular transcription.

The cleavage of host proteins can directly or indirectly lead to the disruption of cellular translation and transcription. Such cleavage events are summarized in Table 3.1. Viral modifications of cellular proteins are not restricted to down regulation of host cell machinery but can also extend to the enhancement of viral IRES-mediated translation, viral RNA synthesis, and evasion of the host antiviral and stress response mechanisms. These topics will be further discussed in the following sections of this chapter.

Use and abuse of host cell functions for viral translation and RNA replication

Use of host factors for IRES-dependent translation and viral RNA synthesis

Since Enteroviruses replicate in the host cytoplasm, and a number of cellular proteins involved in viral replication are predominantly nuclear, Enteroviruses modify the cellular nucleo-cytoplasmic trafficking mechanism to accumulate proteins used for viral replication in the cell cytoplasm. Some of the cellular proteins modified during viral infection that will be discussed in this section include La autoantigen, polypyrimidine tract-binding protein (PTB), poly(rC)-binding protein 2 (PCBP2), and serine/arginine (SR)-rich protein (SRp20) (Table 3.2). These shuttling proteins contain an amino acid sequence known as a nuclear localization signal (NLS) that is recognized by a specific import receptor complex (Görlich and Kutay, 1999). Protein–receptor complexes then relocalize from the cytoplasm to the nucleus through nuclear pore complexes (NPCs) embedded in the nuclear envelope of the host cell. The NPC is made up of nucleoporins (Nups) that contain phenylalanine-glycine repeats necessary for shuttling the protein–receptor complex through the nuclear membrane. During *Enterovirus* infection, the NPC becomes modified when poliovirus or human rhinovirus proteinase 2A cleaves Nup62, Nup98, and Nup153. These cleavage events correlate with proteins accumulating in the cytoplasm and inhibition of import pathways (Belov *et al.*, 2000; Castelló *et al.*, 2009; Fitzgerald *et al.*, 2013; Gustin and Sarnow, 2001; Park *et al.*, 2008; Park *et al.*, 2010; Watters and Palmenberg, 2011). Cleavage of Nups results in loss of the phenylalanine-glycine repeats necessary for the protein–receptor complex docking in the NPC domain during shuttling through the nuclear membrane (Bayliss *et al.*, 2000; Stewart *et al.*, 2001). Pathways that are inhibited during *Enterovirus* infection include the transportin import pathway and K nuclear shuttling (KNS) import pathway. The KNS import pathway mediates the transport of RNA-binding proteins required for *Enterovirus* replication known as heterogeneous nuclear ribonucleoproteins (hnRNPs) (Gustin, 2003; Gustin and Sarnow, 2001; Michael *et al.*, 1997; Pollard *et al.*, 1996). This viral-mediated disruption of nucleo-cytoplasmic trafficking results in accumulation of nuclear proteins in the cytoplasm necessary to enhance viral translation and RNA synthesis.

Host protein La was initially described as an autoantigen found in sera from patients with systemic lupus erythematosus and Sjögren syndrome (Tan, 1989). La is predominantly a nuclear protein in the uninfected cell and plays a role in the maturation of RNA pol III transcripts, due to its ability to bind various RNA structures via its RNA binding domain (Gottlieb and Steitz, 1989; Kenan *et al.*, 1991). During poliovirus or coxsackievirus B3 (CVB3) infection, La becomes relocalized to the cell cytoplasm and interacts with the 5′ NCR of the viral RNA to enhance IRES-mediated viral translation (Meerovitch *et al.*, 1993; Ray and Das, 2002). During the course of poliovirus infection, La is cleaved by viral proteinase 3C but is still able to bind the viral IRES and mediate translation of the viral genome (Shiroki *et al.*, 1999). Previous studies showed that the addition of purified La protein to rabbit reticulocyte lysate enhances translation while inhibiting accumulation of aberrant translation products (Meerovitch *et al.*, 1993; Svitkin *et al.*, 1994). La is only one of the several known IRES trans-acting factors (ITAFs) that interact with the viral IRES to enhance translation of the *Enterovirus* genome. Another nuclear RNA binding protein, nucleolin, has also been shown to interact with both the 5′ and 3′ NCR of poliovirus RNA

Table 3.2 Use and abuse of host cell functions for *Enterovirus* translation and RNA replication. *Enterovirus* proteins alter host proteins to stimulate viral translation and replication. Viral proteinases 2A, 3C, and 3CD can mediate cleavage of host proteins to change their canonical functions to non-canonical activities to aid in viral replication. However, some non-structural proteins, including 2B, 2BC, and 3A, can modify the microenvironment of the cytoplasm to generate replication complexes so that viral RNA synthesis can be carried out

Host protein	Viral protein	Function of host protein	Role in viral replication	References
Nup62, Nup98, Nup153	PV and HRV 2A	Nucleo-cytoplasmic trafficking	Concentrates host proteins in the cytoplasm	Park *et al*. (2008, 2010), Watters and Palmenberg (2011), Castelló *et al*. (2009), Gustin and Sarnow (2001)
La	PV 3C	Maturation of RNA pol III transcripts	Binds IRES for translation	Shiroki *et al*. (1999)
PTB	PV and HRV*14* 3C/3CD	Alternative splicing	Full-length PTB enhances viral translation; cleaved PTB mediates switch to RNA synthesis	Back *et al*. (2002)
PCBP2	PV, CVB3, and HRV 3C/3CD	RNA binding; mRNA stability	Full-length PCBP2 binds SLIV to enhance translation; full-length PCBP2 binds SLI to initiate RNA synthesis; cleaved PCBP2 binds SLI for RNA synthesis	Perera *et al*. (2007), Chase *et al*. (2014)
SRp20	PV 2A	Shuttling RNA binding protein; mRNA splicing	Interacts with PCBP to enhance translation	Fitzgerald *et al*. (2013)
COPII	PV 2B	Protein transport from the ER to Golgi	Forms replication complexes	Rust *et al*. (2001)
GBF1, BIG1/2	PV 3A and 3CD	Guanine nucleotide exchange factors	Activated Arf1 to produce PI4KIIIβfor replication complex formation	Belov *et al*. (2007)
LC3	PV 2BC and 3A	Autophagy	Forms replication complexes	Jackson *et al*. (2005), Taylor and Kirkegaard (2007)
AUF1	PV and HRV 3CD; CVB3 3C	mRNA decay; RNA stability	Negative regulator of translation	Rozovics *et al*. (2012), Wong *et al*. (2013), Catchacrt *et al*. (2013)
TDP2	None	DNA repair	Cleaves VPg from 5′ end	Virgen-Slane *et al*. (2012)

to stimulate viral translation and replication, although the exact mechanism remains unclear (Izumi *et al.*, 2001; Waggoner and Sarnow, 1998).

Host cell shuttling protein PTB is a member of the hnRNP complex and also functions as a cellular ITAF. In the uninfected cell, PTB functions as a repressive regulator of alternative splicing (Mulligan *et al.*, 1992). During poliovirus infection, PTB, like La, relocalizes to the cell cytoplasm and becomes redistributed (Back *et al.*, 2002). Full-length PTB has been implicated in enhancing viral translation during infection (Florez *et al.*, 2005) by binding to stem–loop V of the poliovirus IRES and modulating adjacent eIF4G binding (Kafasla *et al.*, 2010). Full-length PTB also interacts with PCBP2 when bound to stem–loop IV of the poliovirus 5′ NCR to stimulate translation (Kim *et al.*, 2000). However, the multiple isoforms of PTB are cleaved between the RRM domains by poliovirus 3C/3CD proteinase late during viral infection (Back *et al.*, 2002). Cleavage of the N-terminal domain of PTB could result in loss of interaction with PCBP2 and the hnRNP complex. Alternatively, cleavage of the C-terminal domain could result in loss of interaction with the IRES element (Back *et al.*, 2002). The accumulation of cleaved PTB corresponded with a decrease in viral translation levels *in vitro*, suggesting a role in mediating the switch from viral translation to negative-strand RNA synthesis during the replication cycle (Back *et al.*, 2002). A switch in viral translation to RNA synthesis is required during viral replication since the positive-strand viral RNA is translated in a 5′ to 3′ direction by the translation machinery, while the negative-strand viral RNA is synthesized using the same template but in the opposite direction by the virus-encoded RdRP $3D^{pol}$ [reviewed in Daijogo and Semler (2011)]. Interestingly, cleavage of PTB seems to be specific to poliovirus- or human rhinovirus-infected HeLa cells. A recent study shows that PTB is not efficiently cleaved in human rhinovirus-infected WisL cells, a human lung fibroblast cell line, suggesting that host proteins may be differentially cleaved by Enteroviruses in different cell lines (Chase and Semler, 2014). RNA-binding host protein unr has been shown to act synergistically with PTB to enhance human rhinovirus IRES-mediated translation but has minimum enhancement of poliovirus IRES translation (Hunt *et al.*, 1999). This difference in host protein usage among Enteroviruses suggests that these viruses may utilize different cellular proteins to mediate the same viral functions, including the switch in viral translation to RNA replication.

PCBP2 is a host cell RNA-binding protein that functions as an ITAF for *Enterovirus* translation and has been shown to be involved in the switch from viral translation to RNA synthesis. PCBP2 binds to poly(rC) regions of RNA and is expressed in both the nucleus and cytoplasm of the uninfected cell. During poliovirus infection, PCBP2 binds to RNA secondary structure stem–loop IV of the viral IRES, along with host splicing factor SRp20, to help form the hnRNP complex necessary for viral translation (Bedard *et al.*, 2007; Blyn *et al.*, 1996, 1997). PCBP2 can also bind stem–loop I in the 5′ NCR of the poliovirus genome to form a ternary complex with viral precursor proteinase 3CD (Parsley *et al.*, 1997). This complex is required for initiation of negative-strand RNA synthesis and has been suggested to also be involved in positive-strand RNA synthesis (Gamarnik and Andino, 1997; Parsley *et al.*, 1997; Vogt and Andino, 2010). During poliovirus, coxsackievirus, or human rhinovirus infection of HeLa cells, PCBP2 is cleaved in the linker region between its K-homologous (KH) domains, KH2 and KH3, by viral proteinase 3C/3CD (Chase *et al.*, 2014; Perera *et al.*, 2007). Cleaved PCBP2 can no longer bind stem–loop IV or interact with SRp20, resulting in inhibition of IRES-mediated translation, but it can still bind stem–loop I for viral RNA synthesis (Bedard *et al.*, 2007; Chase *et al.*, 2014; Perera *et al.*, 2007). It has also been suggested that poliovirus 3CD binds to stem–loop I to increase the binding affinity of PCBP2 to stem–loop I, thus decreasing its availability for binding to stem–loop IV for translation (Gamarnik and Andino, 1998). Cleavage of PCBP2, along with the cleavage of other proteins such as PTB, can help mediate the switch from viral translation to RNA synthesis. Interestingly, cleavage of PCBP2 does not occur in human rhinovirus-infected human lung fibroblasts, WisL cells, while it does when they are infected with poliovirus, as determined by Western blot analysis (Chase and Semler, 2014). Such a differential cleavage pattern suggests that cleavage of specific host proteins may be required to mediate the switch to RNA synthesis only in certain cell types. It is also possible that the concentration of PCBP2 in WisL cells is low and below the level of detection of the Western blot analysis used in this study (Chase and Semler, 2014). The mechanism that brings about the switch from viral translation to RNA replication remains incompletely understood and will require future studies.

Host protein SRp20, a shuttling protein involved in mRNA splicing and translation, contains an RRM domain at its N-terminus for RNA binding and a serine/arginine (RS)-rich domain at its C-terminus for nucleo-cytoplasmic shuttling and protein–protein interactions (Cáceres *et al.*, 1997, 1998). During poliovirus or CVB3 infection, and to a lesser extent during human rhinovirus 16 infection, SRp20 relocalizes to the infected cell cytoplasm (Fitzgerald *et al.*, 2013; Fitzgerald and Semler, 2011). Poliovirus proteinase 2A activity is required for this redistribution pattern in the cell cytoplasm (Fitzgerald *et al.*, 2013). In the cytoplasm, SRp20 enhances poliovirus translation by binding the KH3 domain of PCBP2 with its RS domain and recruiting ribosomes to the IRES for translation (Bedard *et al.*, 2007). Whether SRp20 recruits the ribosomes directly or indirectly to stem–loop IV for IRES-mediated translation remains to be determined. It is possible that additional undiscovered ITAFs are required to recruit ribosomes to the IRES for translation or that SRp20 may recruit the ribosomes via direct interactions.

Following the initial rounds of translation, viral polyproteins are processed by the virus-encoded proteinases. The non-structural viral proteins can go on to function in viral RNA synthesis. To allow for efficient viral RNA synthesis to occur, there is a switch from viral translation to RNA synthesis. As discussed above, this switch is currently thought to occur when host factors, such as PTB and PCBP2, are cleaved by *Enterovirus* proteinase 3C/3CD. Cleavage of ITAFs inhibits IRES-mediated translation but still allows viral RNA replication to proceed, since the presence of these cleaved proteins favours the clearing of ribosomes from the RNA template. Overall, viral translation and RNA replication are dependent on the modifications of host proteins by the virus-encoded proteinases. In addition to host protein modifications, the cellular environment becomes altered in *Enterovirus*-infected cells so efficient viral RNA replication can occur.

Alteration of host cell membranes for viral RNA synthesis

For viral RNA replication to occur, cellular organelles must be modified to form virus-induced membranous vesicles that serve as sites of replication complexes for viral RNA synthesis (Caliguiri and Tamm, 1969; Dales *et al.*, 1965). The specific localization of these membranous vesicles may physically separate RNA synthesis from IRES-mediated translation in the cytoplasm and increase the local concentrations of viral proteins required for viral replication. The virus-induced vesicles are derived from the endoplasmic reticulum

(ER), Golgi, and from components of autophagic vesicles to form single- and double-walled vesicles (Bienz *et al.*, 1987; Jackson *et al.*, 2005; Schlegel *et al.*, 1996).

The COPII complex components, Sec13 and Sec31, have been shown to colocalize with viral protein 2B, suggesting that COPII may be involved in the formation of replication complexes (Rust *et al.*, 2001). COPII is a vesicle coat protein complex that transports proteins from the ER to the Golgi in the uninfected cell (Barlowe *et al.*, 1994). COPII vesicle proteins are made in the ER with the help of COPII complexes, which include coat proteins. Once the COPII-coated vesicles are formed, they bud from the ER, lose their coat proteins, and become fused to the Golgi (Klumperman, 2000; Rust *et al.*, 2001; Springer *et al.*, 1999). During poliovirus infection, it has been shown that these ER-derived vesicles accumulate in the cytoplasm, and there is a transient increase in COPII vesicle budding from the ER (Bienz *et al.*, 1987; Rust *et al.*, 2001; Trahey *et al.*, 2012). Alterations in the secretory pathway also mediate the formation of viral replication complexes. During poliovirus infection, non-structural viral protein 3A recruits guanine nucleotide exchange factor (GEF), GBF1, and viral proteinase 3CD recruits GEFs, BIG1, and BIG2, to membranes to activate the secretory pathway by converting the small GTPase Arf1 into its GTP-active form (Belov *et al.*, 2007). Arf1-GTP can alter membrane curvature and recruit coat proteins to form secretory transport vesicles (Belov and Ehrenfeld, 2007). The activation of Arf1 leads to the production of phosphatidylinositol-4-phosphate (PI4P), a lipid with an important role in vesicle transport. Expression of CVB3 3A can lead to an accumulation of PI4P and PI4-kinase III β (PI4KIIIβ) (Hsu *et al.*, 2010). During CVB3 infection, PI4KIIIβ has also been shown to colocalize with sites of viral RNA replication and to be required for both poliovirus and CVB3 replication (Hsu *et al.*, 2010). 3A has been shown to associate with acyl coenzyme A [acyl-CoA]-binding protein domain 3 (ACBD3), a protein that binds to an integral Golgi protein known as giantin (Greninger *et al.*, 2012). However, it was recently shown that although ACBD3 does interact with CVB3 3A and PI4KIIIβ directly, this interaction is not required for the recruitment of PI4KIIIβ to replication complexes (Dorobantu *et al.*, 2014). Additionally, depletion of GBF1 and Arf1 by pharmalogical inhibition or small interfering RNA (siRNA) treatment in CVB3-infected cells did not inhibit PI4KIIIβ recruitment (Dorobantu *et al.*, 2014). These contradictory findings suggest that the mechanism for virus-induced host membrane reorganization remains poorly understood and additional studies are required to dissect the involvement of the secretory pathway during *Enterovirus* replication.

Host cell membrane organization throughout poliovirus infection has been observed by electron microscopy (Belov *et al.*, 2012; Caliguiri and Tamm, 1970). At 3 hours post infection (hpi), replication complexes appear to be single-membraned, while at 4 hpi the complexes appear to be convoluted. At later times of infection, replication complexes appear to be double-membraned, illustrating the dynamic nature of viral-induced, membranous vesicles throughout the replication cycle. The convoluted membranes observed at peak times of infection resemble the crescent-shaped precursor membranes seen during autophagy. During poliovirus infection, LC3, a marker for autophagy, localizes to these membranous vesicles. This localization is induced by viral proteins 2BC and 3A (Jackson *et al.*, 2005; Taylor and Kirkegaard, 2007). One hypothesis to explain these observations is that Enteroviruses induce formation of replication complexes via a mechanism similar to autophagy (Kemball *et al.*, 2010; Klein and Jackson, 2011; Suhy *et al.*, 2000). However, a recent study using an antibody specific for double-stranded RNA (dsRNA), which is an

RNA intermediate formed during viral RNA synthesis, to identify replication complexes found that dsRNA does not significantly colocalize with LC3 early during infection but does at late times of infection (Richards *et al.*, 2014). Although this contradicts previous studies suggesting that LC3 plays a role in replication complex formation, the authors of this report alternatively suggest that LC3 may have a role in viral replication, but not in complex formation (Richards *et al.*, 2014). This apparent discrepancy in findings may be due to previous studies using antibodies against viral proteins to analyse the role of the autophagy pathway during viral RNA synthesis instead of antibodies specific for viral RNA replication intermediates. In summary, although previous studies have attempted to elucidate the mechanisms utilized in *Enterovirus* replication complex formation, there are many features of this process that remain to be determined.

Additional host proteins usurped for viral translation and replication

Enteroviruses require multiple host factors to carry out their viral replication cycles. It is apparent that the host factors described above are not sufficient to carry out translation, replication, and encapsidation of the viral RNA. In a comprehensive attempt to identify host factors binding to the viral RNA during infection, numerous experimental approaches have been employed. A recent study using thiouracil cross-linking mass spectrometry (TUX-MS) has identified host factors binding to poliovirus RNA during replication in HeLa cells (Lenarcic *et al.*, 2013). In addition to previously identified host factors known to interact with the viral RNA, 66 putative host proteins have been identified using this methodology. From these 66, eight proteins were selected for validation. Knockdown of two of these proteins, NONO (non-POU-domain-containing octamer-binding protein) and CNBP (cellular nucleic acid-binding protein), decreased poliovirus replication similar to levels when PCBP2, La, PTB, or hnRNP C was knocked down. Further analysis of these two host proteins revealed that CNBP was required for efficient viral translation and NONO was required for efficient positive-strand RNA synthesis (Lenarcic *et al.*, 2013). This methodology proved to be an effective way to identify proteins associated with the viral RNA during *Enterovirus* replication. One of these proteins is AU-rich binding factor 1 (AUF1) (Lenarcic *et al.*, 2013), which had been previously identified via an RNA affinity screen for proteins interacting with the 5′ NCR (Rozovics *et al.*, 2012).

AUF1, also known as hnRNP D, is a cellular protein that binds to AU-rich elements in the 3′ NCR of mRNAs in the uninfected cell (Zhang *et al.*, 1993). It is involved in RNA stability and can target RNAs for degradation via an mRNA-decay pathway (Kiledjian *et al.*, 1997). AUF1 has four isoforms due to alternative splicing that contain tandem RRMs to bind RNA (Kajita *et al.*, 1995). During poliovirus and CVB3 infection, AUF1 relocalizes from the nucleus to the cytoplasm in a proteinase 2A-driven manner and colocalizes with the non-structural viral protein 2B (Cathcart *et al.*, 2013). AUF1 is cleaved by poliovirus or human rhinovirus 3CD and CVB3 3C (Rozovics *et al.*, 2012; Wong *et al.*, 2013). AUF1 has also been shown to directly interact with the poliovirus 5′ NCR, specifically full-length 5′ NCR and stem–loop IV (Cathcart *et al.*, 2013; Rozovics *et al.*, 2012). This interaction is inhibited by the cleavage of AUF1 by 3CD (Cathcart *et al.*, 2013). AUF1 has been shown to interact with the 3′ NCR of CVB3 RNA as well, due to the AU-rich sequence at the 3′ end (Wong *et al.*, 2013). When AUF1 is genetically ablated or knocked down, poliovirus, human rhinovirus 16, or CVB3 viral titres increase, suggesting an inhibitory role for AUF1 during *Enterovirus* infection (Cathcart *et al.*, 2013; Wong *et al.*, 2013). AUF1 has been shown to decrease poliovirus translation *in vitro*, suggesting that AUF1 functions as an antiviral factor during *Enterovirus* infection (Cathcart *et al.*, 2013). It is possible that Enteroviruses cleave AUF1 to disrupt the interaction of this protein with viral RNA as a mechanism to evade the cellular RNA decay pathway. AUF1 has also been shown to interact with other host factors involved in viral replication, including PCBP2, nucleolin, and PABP (Dempsey *et al.*, 1998; Kiledjian *et al.*, 1997; Lu *et al.*, 2006). AUF1 cleavage may disrupt these host protein–protein interactions so that these host factors can bind to the viral RNA and stimulate viral translation and replication. Another possible role for AUF1 during *Enterovirus* replication stems from the fact that AUF1 has been shown to bind to both the 5′ and 3′ NCRs of the viral RNA. Such interactions could aid in the functional circularization of viral RNA through homo-multimerization of AUF1 via its dimerization domain present at the N-terminus. Further studies are necessary to define the role of AUF1 during *Enterovirus* infection and how the virus might evade the RNA degradation pathway initiated by this protein.

Another recently identified host protein that is utilized during *Enterovirus* replication is 5′ tyrosyl-DNA phosphodiesterase 2 (TDP2) (Virgen-Slane *et al.*, 2012). In the uninfected cell, TDP2, also known as TRAF and TNF receptor-associated protein (TTRAP) and ETS-1 associated protein II (EAPII), functions as a DNA repair enzyme that hydrolyses 5′ phosphotyrosyl DNA linkages in topoisomerase II-mediated double-strand breaks (Cortes Ledesma *et al.*, 2009). TDP2 was shown to harbour the activity, discovered decades ago, that cleaves the phosphotyrosyl bond between VPg and the 5′ end of poliovirus virion RNA (Ambros and Baltimore, 1978; Ambros *et al.*, 1978; Virgen-Slane *et al.*, 2012). This activity was initially referred to as VPg unlinkase and was shown to be present in both the nucleus and the cytoplasm of uninfected and poliovirus-infected cells (Ambros *et al.*, 1978). TDP2 is a predominantly nuclear protein but is also found in the uninfected cell cytoplasm. During poliovirus infection, TDP2 relocalizes from the nucleus to the cell cytoplasm, and during peak times of infection, is sequestered to the cell periphery in sites distinct from putative replication complexes (Virgen-Slane *et al.*, 2012). Earlier studies had shown that VPg is absent from actively translating viral RNAs but present on newly synthesized viral RNAs or encapsidated virion RNAs (Fernandez-Munoz and Darnell, 1976; Hewlett *et al.*, 1976; Nomoto *et al.*, 1977; Nomoto *et al.*, 1976). However, another early study reported that poliovirus RNA with a VPg linked to the 5′ end was capable of forming a translation initiation complex in rabbit reticulocyte lysate (RRL), suggesting that the cleavage of VPg may not be required for viral translation *in vitro* (Golini *et al.*, 1980). Limitations to the experiments in this latter study include that they were done in a cell-free system using RRL, a lysate deficient in TDP2 (Rozovics *et al.*, 2011), and that the sucrose gradients could not determine if the VPg linked RNA represented a small or larger population of the viral RNA associated with the ribosomes.

In accordance with the early findings described above, a study using a replicon with an uncleavable bond between VPg and the viral RNA and a reporter gene shows that following transfection of this mutated CVB3 or poliovirus RNA, viral translation and replication are unaltered (Langereis *et al.*, 2014). Although these results suggest that cleavage of the VPg-RNA linkage is not required for viral translation and replication, the caveats of the study must also be considered. The viral RNA harbouring this uncleavable bond and reporter gene was transfected into the cells, thus bypassing the normal receptor-mediated entry pathway and the uncoating step. These steps that occur during normal infection may be essential in determining the orientation of viral RNA during uncoating and its initial exposure to the cell cytoplasm following uncoating. It is possible that since the viral RNA has been transfected into the cell and thus only undergoes primary rounds of translation and replication, the cleavage of VPg by VPg unlinkase/TDP2 may be necessary to determine the fate of nascent viral RNAs to either be encapsidated in progeny virus particles or undergo an additional round of translation and replication. A recent study showed that TDP2 is necessary for efficient *Enterovirus* replication in murine cells (Maciejewski *et al.*, 2015). Although viral yields were significantly reduced in the absence of TDP2, this study shows that TDP2 is differentially required among Enteroviruses. The greatest dependency on TDP2 was seen during CVB3 infection, with CVB3 yields remaining unchanged in the absence of TDP2. However, in agreement with Langereis *et al.* (2014), CVB3 replicated at reduced levels in the absence of TDP2 following transfection using a CVB3 replicon encoding a reporter gene (Maciejewski *et al.*, 2015). As stated above, it is important to note the experimental caveats of performing a transfection versus infection. Together these results show that TDP2 plays a critical role during the course of *Enterovirus* infections and could serve as a putative target for antiviral development.

Most studies characterizing host proteins required for the *Enterovirus* replication cycle have been carried out using the prototypic *Enterovirus*, poliovirus. As discussed in this section, not all Enteroviruses require the same host proteins for their viral replication cycles. This variation leaves room for other host proteins to be utilized during replication. Since there have been only a few reports of ITAFs required for EV71, a study was undertaken to identify cellular proteins bound to a biotinylated EV71 5′NCR. From this study, 12 cellular proteins were identified to interact with the 5′ NCR (Lin *et al.*, 2008). Of these 12 proteins, previously identified proteins were purified, including PTB, poly(C)-binding protein 1 (PCBP1, also known as hnRNP E1), PCBP2, La, and Unr. In addition, proteins previously unidentified as EV71 5′ NCR-binding proteins were reported, including hnRNP K, hnRNP A1, far-upstream element-binding protein 1 (FBP1), and FBP2 (Lin *et al.*, 2008). More recently, these latter proteins have been shown to redistribute from the cell nucleus to the cytoplasm and stimulate EV71 infection (Huang *et al.*, 2011; Lin *et al.*, 2008, 2009). Although the roles of hnRNP K, hnRNP A1, FBP1, and FBP2 during enterovirus 71 infection remain to be determined, this finding suggests that although Enteroviruses have a highly conserved genome, they utilize several different host factors to enhance their viral replication.

Evasion of host antiviral and stress response pathways and mRNA surveillance

Enteroviruses, like any invading pathogen, elicit immune and stress responses in the host. This leads to the activation of pathways that induce the interferon response (IFN) and stress granule (SG) formation. Similar to the *Enterovirus*-mediated modifications of host proteins involved in cellular translation and transcription, Enteroviruses can also modify host proteins involved in antiviral and stress response pathways. Additionally, Enteroviruses can evade cellular mRNA surveillance by disrupting processing bodies (P bodies). This section will discuss how Enteroviruses modify host proteins involved in cellular defence pathways to suppress their activation and, thereby, stimulate viral replication.

RIG-I-like receptor detection of Enteroviruses

The host must be able to recognize invading pathogens to avoid eliciting a response against itself. To accomplish this, cells have specialized pattern recognition receptors (PRRs) to detect pathogen-associated molecular patterns (PAMPs). PRRs fall into several different families and include the Toll-like receptors (TLRs), NOD-like receptors (NLRs), C-Type lectin receptors (CLRs), and RIG-like receptors (RLRs). Two essential immune receptors in the RLR family that detect double stranded viral RNA in the cytoplasm are retinoic acid-inducible gene I (RIG-I) and melanoma differentiation-associated protein 5 (MDA5). Ligand binding causes these receptors to oligomerize and serve as signalling platforms to recruit and activate signalling adaptors, including the mitochondrial adaptor, MAVS. This activates a cascading signalling pathway that leads to the transcription of NF-κB factors and interferon regulatory factors (IRFs). These transcription factors then go on to induce expression of type I IFN. Such cellular response pathways induced by the virus serve as the front line of host defence against invading pathogens (reviewed in Reikine *et al.*, 2014).

RLRs belong to the DExD/H-box helicase family and use ATP binding and hydrolysis to drive formation of an oligomer filament composed of repeating dimers, which extends from their target protein to downstream signals to activate IFN signalling (Peisley *et al.*, 2013). The DExD/H-box helicase family includes three sensors for detecting viruses: RIG-I, MDA5, and LGP2. RIG-I discriminates between host and viral RNA by specifically binding to short double-stranded RNAs (dsRNAs) and single-stranded RNAs (ssRNAs) (less than 300 bp) with a 5′ triphosphate (5′ppp) moiety (Hornung *et al.*, 2006). The C-terminal domain of RIG-I recognizes and binds 5′ppp, while its helicase domains bind the RNA to form oligomer filaments (Hornung *et al.*, 2006; Peisley *et al.*, 2013). In contrast to RIG-I, MDA5 binds specifically to long dsRNA regardless of its 5′ end moiety. MDA5 also forms a filament around dsRNA. Both RIG-I and MDA5 contain an N-terminal caspase recruitment domain (CARD), which aids in MAVS recruitment and activation (Kato *et al.*, 2006). However, LGP2, the third member of the family, lacks a CARD and therefore cannot recruit or activate MAVS. Instead, LGP2 responds to viral stimuli by modulating RIG-I and MDA5 signalling (Childs *et al.*, 2013). Since *Enterovirus* RNAs lack the 5′ppp moiety, the interaction between RIG-I and *Enterovirus* RNAs is not well understood but will be discussed later in this section. To date, LGP2 has not been shown to interact with *Enterovirus* RNAs but has been recently shown to interact with the RNA of encephalomyocarditis virus, another member of the *Picornaviridae* family belonging to the *Cardiovirus* genus (Deddouche *et al.*, 2014). However, MDA5 along with its adaptor molecule MAVS, has been shown to induce the antiviral response pathway during *Enterovirus* infection and thus will be the focus of this section.

MDA5 serves as a cytoplasmic sensor for *Enterovirus* infections (Abe *et al.*, 2012; Wang *et al.*, 2010). MDA5 was shown to specifically bind to two dsRNA *Enterovirus* species, the replicative form (RF) and the replicative intermediate (RI) (Feng *et al.*, 2012; Triantafilou *et al.*, 2012). The RF is formed when the negative-strand is being synthesized from the positive-strand template, and the RI is formed when there are multiple positive-strands being synthesized from the negative-strand [reviewed in Daijogo and Semler (2011) and Baltimore *et al.* (1964)]. These two viral species could induce type I IFN response following transfection of *Enterovirus* RNA into the cell (Feng *et al.*, 2012; Triantafilou *et al.*, 2012). Gel-purified CVB3 RF alone was also shown to activate the ATPase activity of recombinant MDA5 (Feng *et al.*, 2012). Interestingly, transfecting CVB3 genomic RNA into mouse embryonic fibroblasts knocked out for RIG-I did not significantly alter its ability to produce an IFN response (Feng *et al.*, 2012). These results are not surprising since RIG-I requires a 5′ppp moiety in short dsRNA to interact with the target RNA. When mice lacking MDA5 and its adaptor molecule MAVS were infected with CVB3, there was an increase in mortality and a reduced IFN response (Wang *et al.*, 2010). However, poliovirus-infected mice expressing the poliovirus receptor but lacking MDA5 and MAVS did not display increased mortality nor a decrease in IFN response even though it was shown that MDA5 was required for the production of IFNs *in vitro* (Abe *et al.*, 2012; Oshiumi *et al.*, 2011). It is important to note that the apparent discrepancy seen between the *in vitro* versus *in vivo* data may be due to differences in the viral RNA delivery method, host protein association with the viral RNA, or overall physiological environment of the cell.

A recent study was carried out to measure the type I IFN response, specifically IFN-α/β, at the different steps of the CVB3 replication cycle. When negative-strand RNA synthesis was inhibited, thereby inhibiting the production of RF, the IFN-α/β response was abolished; when positive-strand RNA synthesis was inhibited, resulting in inhibition of the production of RI, the IFN-α/β response was not impaired (Feng *et al.*, 2012). These results suggest that RF is required to activate the type I IFN response during an *Enterovirus* infection. Also, MDA5 and MAVS are required to sense the dsRNA intermediates produced during *Enterovirus* infection. To evade the IFN response pathway, Enteroviruses modify the RLRs directly as well as modify the host proteins downstream of the RLRs signalling pathway.

Enterovirus evasion of the RIG-I-like receptor-mediated antiviral response

Enteroviruses employ different mechanisms to disrupt the type I IFN pathway. Typically, they modify host proteins in a proteinase-dependent manner. However, there are differing reports for virus-mediated modifications of the antiviral response among Enteroviruses. *Poliovirus* has been shown to degrade MDA5 and MAVS in a proteasome- and caspase-dependent manner (Barral *et al.*, 2007; Rebsamen *et al.*, 2008). CVB3 proteinase 3C cleaves MAVS, resulting in a functionally inactive fragment that can no longer participate in both NF-κB and type I IFN signalling (Mukherjee *et al.*, 2011). CVB3 has also been shown to target other components of the RLR pathway, including 3C-mediated cleavage of focal adhesion kinase (FAK), which is responsible for MAVS induction (Bozym *et al.*, 2012). For EV71, proteinase 2A cleaves MAVS (Wang *et al.*, 2013). EV71 3C can either directly cleave a downstream signalling protein, interferon regulatory factor 7 (IRF7), or block the recruitment of MAVS to RIG-I, thus interrupting the relocalization of downstream signalling protein IRF3 to the nucleus and reducing the IFN response (Lei *et al.*, 2010, 2013). Similar to poliovirus, EV71 also induces caspase-mediated degradation of MDA5 (Kuo *et al.*, 2013). Human rhinovirus type 1a proteinases 2A and 3C and cellular caspase 3 have all been suggested to cleave MAVS (Drahos and Racaniello, 2009).

Recently a comprehensive study was carried out to analyse the viral-mediated cleavage of key players involved in the RLR-mediated IFN response. In this study, *Enterovirus* modulation of the major components of the RLR signalling pathway was studied. CVB3 proteinase 2A was shown to cleave both MDA5 and MAVS (Feng *et al.*, 2014). MDA5 and MAVS cleavage was also observed when using recombinant EV71, CVB3, and poliovirus 2A proteinases, suggesting that Enteroviruses uniformly utilize proteinase 2A to cleave MDA5 and MAVS and thus inhibit the type I IFN response (Feng *et al.*, 2014) (Fig. 3.2). The apparent

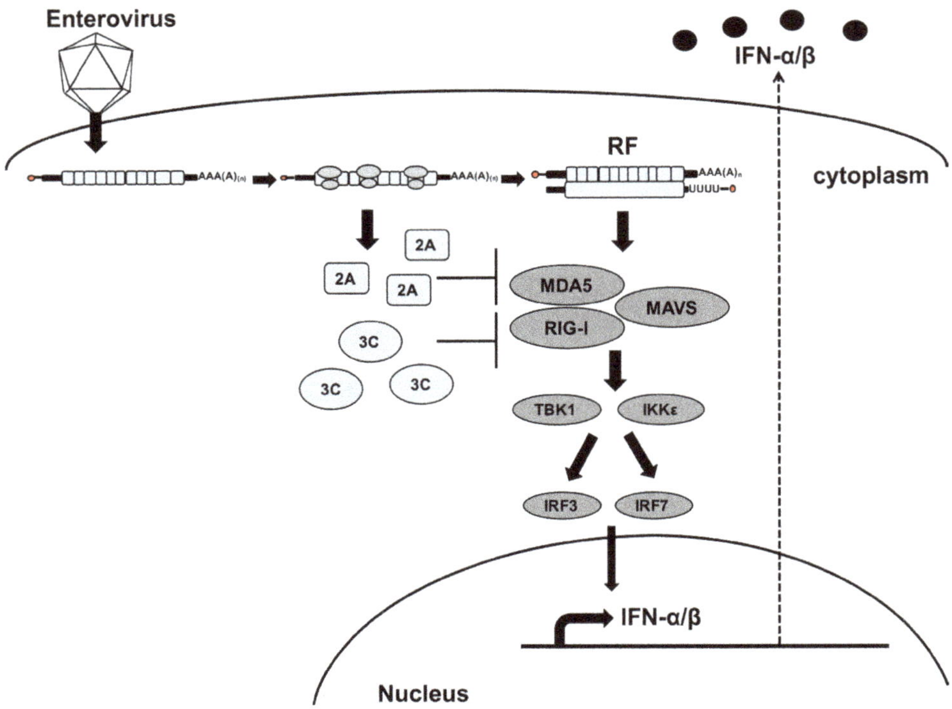

Figure 3.2 *Enterovirus* evasion of the type I IFN response. The host elicits an immune response when it detects invading pathogens. The immune receptors of the RLR pathway (RIG-I and MDA5) can detect double stranded RNAs. Although Enteroviruses have single stranded RNAs, a double stranded replicative form (RF) is generated during viral replication. The RLR pathway is induced by detection of this RF. In order to bring about efficient replication of their viral genomes, Enteroviruses have evolved mechanisms to evade the host antiviral response. A recent comprehensive study has shown that *Enterovirus* proteinase 2A can cleave both MDA5 and MAVs and proteinase 3C can cleave RIG-I to inhibit the type I IFN response (Feng *et al.*, 2014).

differences between these findings and previous reports may be because in previous studies, cells were pre-treated before infection with cell stress inducer, polyI:C, which may have activated caspase-mediated apoptosis. The previous study (Mukherjee *et al.*, 2011) showing 3C-mediated cleavage of MAVS was carried out using exogenous 3C versus the recent study, which used endogenous 3C (Feng *et al.*, 2014). Overexpression of 3C may have preferentially cleaved MAVS. Although it is possible that both 2A and 3C can cleave MAVS, it is likely that during a normal infection 2A preferentially cleaves MAVS (Feng *et al.*, 2014). Additionally, RIG-I is cleaved by recombinant CVB3 3C (Feng *et al.*, 2014). It was unknown if RIG-I recognized Enteroviruses, since previous studies have shown that the type I IFN response is unaltered when RIG-I is knocked out; however, the newly identified CVB3 3C-mediated cleavage event may help elucidate the mechanisms of *Enterovirus* evasion of the RIG-I pathway (Feng *et al.*, 2012). This function may be redundant, since Enteroviruses cleave MDA5 and MAVS to inhibit the IFN response pathway. Perhaps Enteroviruses modify multiple pathways to ensure efficient replication of their genomic RNAs. Cleavage events in the RLR pathway allow the virus to shut down the host antiviral response and efficiently replicate in the cell cytoplasm. Given that Enteroviruses use similar replication mechanisms due to their conserved genomic RNA sequences, it is likely that they mediate similar cleavage events to evade the type I IFN response.

Enterovirus evasion of the stress response

Cells form stress granules (SGs) in the presence of environmental stress stimuli. In the uninfected cell, SGs form cytosolic aggregates that contain cap-dependent stalled pre-initiation complexes, translationally silenced mRNAs, and canonical translation eIFs. These aggregates form following the phosphorylation of eukaryotic initiation factor 2α (eIF2α), which is a required component of the eIF2 complex responsible for loading the tRNA onto the 40S ribosome (Ernst *et al.*, 1978). Although the exact mechanism of SG generation is not yet understood, SGs typically form as a result of hypoxia, nutrient deprivation, or oxidative stress. SG formation is canonically used to halt cap-dependent translation. Host proteins proposed to be important during SG formation include the RNA binding proteins, T-cell restricted intracellular antigen 1 (TIA-1), TIA-1-related protein (TIAR), and RasGAP-SH3 domain-binding protein (G3BP) (Kedersha *et al.*, 1999; Tourrière *et al.*, 2003). Previous studies have shown that SG formation can reduce the levels of *Enterovirus* replication. It has been suggested that viral RNAs are sequestered by SGs or that the host proteins required for viral replication can become trapped in SGs and are thus unavailable for viral replication (White *et al.*, 2007). However, the exact role SGs play during *Enterovirus* infection remains poorly understood.

SGs can form independently of eIF2α phosphorylation. Poliovirus infection can lead to stress granule formation as a stress response from the host (Mazroui *et al.*, 2006). This stress granule formation occurs at the same time eIF4G is cleaved, suggesting that *poliovirus*-mediated cleavage of eIF4G, leading to shut down of cap-dependent translation, may induce SG formation (Mazroui *et al.*, 2006). SG formation induced by poliovirus is then disrupted by viral proteinase 3C as infection proceeds (White *et al.*, 2007). 3C can cleave G3BP, leading to SGs dispersing at a later time in infection. When a recombinant, non-cleavable G3BP was introduced into the cell, SG formation was restored and viral replication was reduced (White *et al.*, 2007). This viral-mediated modification of G3BP, a major component in SG formation, may be one mechanism that Enteroviruses use to evade the stress response. However, another study showed that TIA-1-dependent SGs, in the presence of low levels of G3BP and eIF4G, may still form at a later time in poliovirus infection (Piotrowska *et al.*, 2010).

As mentioned above, SGs can be formed due to different stress stimuli. The presence of an RNA-binding protein previously shown to interact with poliovirus RNA-dependent RNA polymerase $3D^{pol}$, Sam68, was used as a marker for SGs unique to poliovirus infection (McBride *et al.*, 1996; Piotrowska *et al.*, 2010). Sam68 localized with SGs induced during poliovirus infection but not with SGs formed due to heat shock or arsenite stress (Piotrowska *et al.*, 2010). This finding suggests that SGs observed during poliovirus infection are unique to stress induced directly by the virus (Piotrowska *et al.*, 2010). Another study showed that TIA-1-positive SGs did form early during poliovirus infection, but as infection proceeded, TIA-1-positive SGs lacking translation factors, RNA-binding proteins, and mRNA were observed (White and Lloyd, 2011). This type of SG observed later in infection suggests that poliovirus most likely disrupts SGs to release any sequestered translation factors to aid in its replication.

During CVB3 infection, SGs are formed early during infection but are disassembled at 5 hpi. CVB3 3C can also cleave G3BP, suggesting a similar mechanism of stress response evasion to poliovirus (Fung *et al.*, 2013). These findings suggest that the viral-mediated modification of the stress response may be conserved among Enteroviruses. Although the exact role of SG formation during *Enterovirus* infection remains unclear, another study reported that host factor, SRp20, which was previously discussed in Table 3.2, colocalizes with SG marker TIA-1 at 3 hpi during a poliovirus infection of HeLa cells. Although not all SRp20 was present in SGs in these experiments, colocalization was unique to poliovirus-induced SG formation, as this colocalization was not observed during oxidative stress (Fitzgerald and Semler, 2013). These results further confirm that translation factors involved during viral replication may be sequestered into SGs and suggest that perhaps viral-mediated disruption of SGs must be achieved to release factors necessary for efficient viral replication.

Evasion of cellular mRNA surveillance

P bodies are another type of cytoplasmic granules that contain translationally silenced, deadenylated mRNAs and may serve in mRNA surveillance of the uninfected cell. P bodies have been suggested to be involved in mRNA decay pathways and include proteins involved in the formation of mRNA decapping complexes, such as Dcp1a/2 and Edc3, major 5′ exonuclease Xrn1, and proteins involved in mRNA deadenylation, Ccr4 and Pan 2/3 (reviewed in Adjibade and Mazroui, 2014). P bodies form aggregates in a similar fashion to SGs, although the mechanism of their formation remains unknown. Since Enteroviruses have positive-strand RNAs lacking a cap at the 5′ end, this makes them susceptible to the mRNA decay pathways. One study demonstrated that P bodies are disrupted at 4 hpi during poliovirus or CVB3 infection (Dougherty *et al.*, 2011). This dispersion of P bodies correlated with an increase of poliovirus 3C/3CD expression. During poliovirus infection, the number of P bodies was shown to decrease in the presence of the viral RNA replication inhibitor, guanidine HCl; however, the P bodies that did not disperse appeared to increase in size (Dougherty *et al.*, 2011). P body protein components Xrn1, Dcp1a, and Pan 3 were also shown to undergo accelerated proteasome-mediated degradation during poliovirus infection. Only Dcp1a appeared to be cleaved by poliovirus proteinase 3C. An increase in turnover of Dcp1a during infection was observed, suggesting that poliovirus targets the mRNA decay pathway to disrupt P body formation (Dougherty *et al.*, 2011). Although the mechanism of *Enterovirus* disruption of P body formation remains to be discovered, it is likely that Enteroviruses disrupt this host function to avoid the mRNA decay pathway.

Summary

Due to their limited coding capacity, Enteroviruses hijack host cell functions to stimulate viral translation and replication (Fig. 3.3). They typically do this by cleaving cellular proteins to modify their canonical functions. Enteroviruses modify proteins involved in cellular nucleo-cytoplasmic trafficking, translation, and transcription to make these proteins available for viral translation and RNA replication. The viral proteinases can also further modify specific host proteins to mediate the switch from viral translation to RNA synthesis. Non-structural viral proteins without proteolytic activity also alter cellular functions, such as membrane reorganization for viral replication complex formation. Additionally, Enteroviruses can evade the host antiviral or stress response to ensure efficient replication. Although

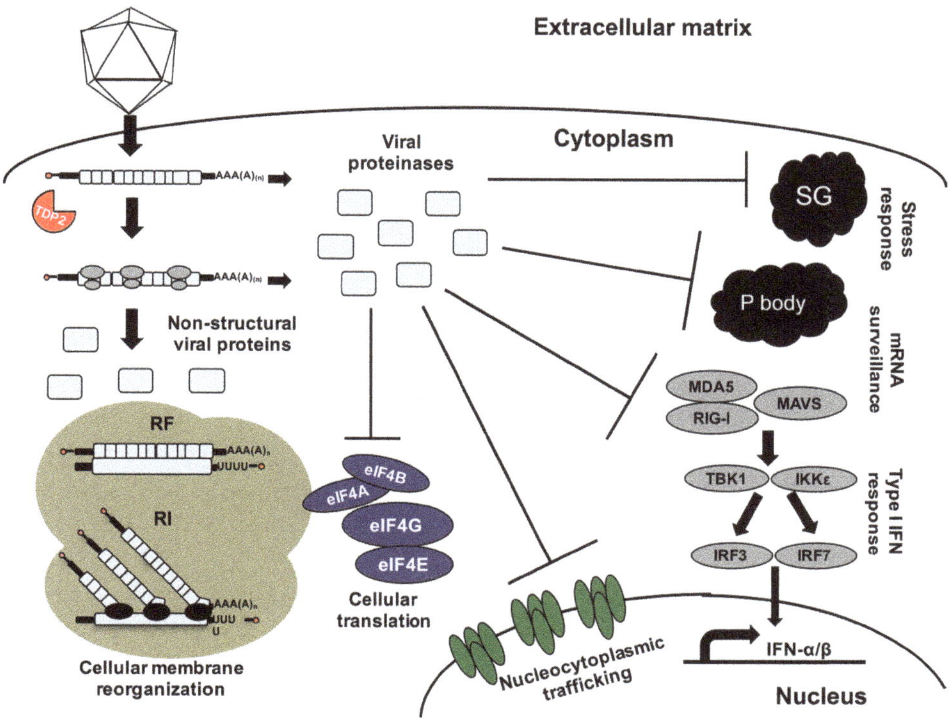

Figure 3.3 Summary of *Enterovirus*-mediated host modifications to enhance viral translation and RNA replication. Enteroviruses modify cellular functions to stimulate viral replication. Upon release into the cytoplasm, the positive-strand viral RNA is translated. It has been suggested that cellular enzyme TDP2 cleaves the covalently linked viral protein, VPg, from the 5′ end of the RNA to allow for polysome association. Following translation of the viral genome, the viral polyprotein is proteolytically processed. The viral proteins can then alter a number of cellular proteins, resulting in the hijacking of host functions for viral replication. Viral proteinases 2A, 3C, and 3CD are responsible for cleaving host proteins involved in cellular nucleo-cytoplasmic trafficking, translation, transcription, and the antiviral response. Non-structural viral proteins, such as 2B, 2BC, and 3A, are also responsible for inducing conformational changes in the host cytoplasmic membranes to serve as replication sites for viral RNA synthesis. Together these virus-induced modifications of cellular proteins result in an altered microenvironment that allows the virus to replicate efficiently.

it is known that Enteroviruses can modify the host functions in multiple ways, many of the mechanisms remain unclear. It is necessary to elucidate these mechanisms and identify viral specific protein–protein interactions so that antivirals may be generated targeting either the host proteins or viral proteins involved. Although antivirals against cellular proteins can be potentially toxic to the cell, targeting a non-canonical function of a host protein or a novel protein–protein interface may circumvent such an issue. Importantly, identifying cellular targets required for *Enterovirus* replication avoids the issues of generating resistant viral variants due to the high mutation rates of viral RNA-dependent RNA polymerases. Such a prospect holds considerable promise for development of broad-spectrum antiviral therapies to treat *Enterovirus* infections.

Acknowledgements

We are grateful to Nicolas Lévêque, Dylan Flather, Ilya Belalov, Eric Baggs, and Wendy Ullmer for critical comments on the manuscript. Research described from the authors' laboratory was supported by Public Health Service grants AI022693, AI026765, and AI110782 from the National Institutes of Health (to B.L.S.). This material is based upon work supported by the National Science Foundation Graduate Research Fellowship Program under Grant No. (DGE-1321846). Any opinion, findings, and conclusions or recommendations expressed in this material are those of the authors and do not necessarily reflect the views of the National Science Foundation.

References

Abe, Y., Fujii, K., Nagata, N., Takeuchi, O., Akira, S., Oshiumi, H., Matsumoto, M., Seya, T., and Koike, S. (2012). The toll-like receptor 3-mediated antiviral response is important for protection against poliovirus infection in poliovirus receptor transgenic mice. J. Virol. *86*, 185-194. https://doi.org/10.1128/JVI.05245-11

Adjibade, P., and Mazroui, R. (2014). Control of mRNA turnover: implication of cytoplasmic RNA granules. Semin. Cell Dev. Biol. *34*, 15–23. https://doi.org/10.1016/j.semcdb.2014.05.013

Ambros, V., and Baltimore, D. (1978). Protein is linked to the 5′ end of poliovirus RNA by a phosphodiester linkage to tyrosine. J. Biol. Chem. *253*, 5263–5266.

Ambros, V., Pettersson, R.F., and Baltimore, D. (1978). An enzymatic activity in uninfected cells that cleaves the linkage between poliovirion RNA and the 5′ terminal protein. Cell *15*, 1439–1446.

Back, S.H., Kim, Y.K., Kim, W.J., Cho, S., Oh, H.R., Kim, J.E., and Jang, S.K. (2002). Translation of polioviral mRNA is inhibited by cleavage of polypyrimidine tract-binding proteins executed by polioviral 3C(pro). J. Virol. *76*, 2529–2542.

Baltimore, D., Becker, Y., and Darnell, J.E. (1964). Virus-specific double-stranded RNA in poliovirus-infected cells. Science *143*, 1034–1036.

Banerjee, R., Weidman, M.K., Navarro, S., Comai, L., and Dasgupta, A. (2005). Modifications of both selectivity factor and upstream binding factor contribute to poliovirus-mediated inhibition of RNA polymerase I transcription. J. Gen. Virol. *86*, 2315-2322.

Barlowe, C., Orci, L., Yeung, T., Hosobuchi, M., Hamamoto, S., Salama, N., Rexach, M.F., Ravazzola, M., Amherdt, M., and Schekman, R. (1994). COPII: a membrane coat formed by Sec proteins that drive vesicle budding from the endoplasmic reticulum. Cell 77, 895-907.

Barral, P.M., Morrison, J.M., Drahos, J., Gupta, P., Sarkar, D., Fisher, P.B., and Racaniello, V.R. (2007). MDA-5 is cleaved in poliovirus-infected cells. J. Virol. *81*, 3677–3684.

Bayliss, R., Littlewood, T., and Stewart, M. (2000). Structural basis for the interaction between FxFG nucleoporin repeats and importin-beta in nuclear trafficking. Cell *102*, 99–108.

Bedard, K.M., and Semler, B.L. (2004). Regulation of picornavirus gene expression. Microbes Infect. *6*, 702–713. https://doi.org/10.1016/j.micinf.2004.03.001

Bedard, K.M., Daijogo, S., and Semler, B.L. (2007). A nucleo-cytoplasmic SR protein functions in viral IRES-mediated translation initiation. EMBO J. *26*, 459–467.

Belov, G.A., and Ehrenfeld, E. (2007). Involvement of cellular membrane traffic proteins in poliovirus replication. Cell Cycle *6*, 36–38.

Belov, G.A., Evstafieva, A.G., Rubtsov, Y.P., Mikitas, O.V., Vartapetian, A.B., and Agol, V.I. (2000). Early alteration of nucleocytoplasmic traffic induced by some RNA viruses. Virology *275*, 244–248. https://doi.org/10.1006/viro.2000.0427

Belov, G.A., Habbersett, C., Franco, D., and Ehrenfeld, E. (2007). Activation of cellular Arf GTPases by poliovirus protein 3CD correlates with virus replication. J. Virol. *81*, 9259–9267.

Belov, G.A., Nair, V., Hansen, B.T., Hoyt, F.H., Fischer, E.R., and Ehrenfeld, E. (2012). Complex dynamic development of poliovirus membranous replication complexes. J. Virol. *86*, 302–312. https://doi.org/10.1128/JVI.05937-11

Bienz, K., Egger, D., and Pasamontes, L. (1987). Association of polioviral proteins of the P2 genomic region with the viral replication complex and virus-induced membrane synthesis as visualized by electron microscopic immunocytochemistry and autoradiography. Virology *160*, 220-226.

Blair, W.S., and Semler, B.L. (1991). Role for the P4 amino acid residue in substrate utilization by the poliovirus 3CD proteinase. J. Virol. *65*, 6111–6123.

Blyn, L.B., Swiderek, K.M., Richards, O., Stahl, D.C., Semler, B.L., and Ehrenfeld, E. (1996). Poly(rC) binding protein 2 binds to stem-loop IV of the poliovirus RNA 5′ noncoding region: identification by automated liquid chromatography-tandem mass spectrometry. Proc. Natl. Acad. Sci. U.S.A. *93*, 11115-11120.

Blyn, L.B., Towner, J.S., Semler, B.L., and Ehrenfeld, E. (1997). Requirement of poly(rC) binding protein 2 for translation of poliovirus RNA. J. Virol. *71*, 6243–6246.

Bovee, M.L., Lamphear, B.J., Rhoads, R.E., and Lloyd, R.E. (1998). Direct cleavage of elF4G by poliovirus 2A protease is inefficient in vitro. Virology *245*, 241–249.

Bozym, R.A., Delorme-Axford, E., Harris, K., Morosky, S., Ikizler, M., Dermody, T.S., Sarkar, S.N., and Coyne, C.B. (2012). Focal adhesion kinase is a component of antiviral RIG-I-like receptor signaling. Cell Host Microbe *11*, 153–166. https://doi.org/10.1016/j.chom.2012.01.008

Cáceres, J.F., Misteli, T., Screaton, G.R., Spector, D.L., and Krainer, A.R. (1997). Role of the modular domains of SR proteins in subnuclear localization and alternative splicing specificity. J. Cell Biol. *138*, 225–238.

Cáceres, J.F., Screaton, G.R., and Krainer, A.R. (1998). A specific subset of SR proteins shuttles continuously between the nucleus and the cytoplasm. Genes Dev. *12*, 55–66.

Caliguiri, L.A., and Tamm, I. (1969). Membranous structures associated with translation and transcription of poliovirus RNA. Science *166*, 885–886.

Caliguiri, L.A., and Tamm, I. (1970). The role of cytoplasmic membranes in poliovirus biosynthesis. Virology *42*, 100–111.

Castelló, A., Izquierdo, J.M., Welnowska, E., and Carrasco, L. (2009). RNA nuclear export is blocked by poliovirus 2A protease and is concomitant with nucleoporin cleavage. J. Cell. Sci. *122*, 3799-3809. https://doi.org/10.1242/jcs.055988

Cathcart, A.L., Rozovics, J.M., and Semler, B.L. (2013). Cellular mRNA decay protein AUF1 negatively regulates enterovirus and human rhinovirus infections. J. Virol. *87*, 10423–10434. https://doi.org/10.1128/JVI.01049-13

Chase, A.J., and Semler, B.L. (2014). Differential cleavage of IRES trans-acting factors (ITAFs) in cells infected by human rhinovirus. Virology *449*, 35–44. https://doi.org/10.1016/j.virol.2013.10.030

Chase, A.J., Daijogo, S., and Semler, B.L. (2014). Inhibition of poliovirus-induced cleavage of cellular protein PCBP2 reduces the levels of viral RNA replication. J. Virol. *88*, 3192–3201. https://doi.org/10.1128/JVI.02503-13

Childs, K.S., Randall, R.E., and Goodbourn, S. (2013). LGP2 plays a critical role in sensitizing mda-5 to activation by double-stranded RNA. PLOS ONE *8*, e64202. https://doi.org/10.1371/journal.pone.0064202

Clark, M.E., Hämmerle, T., Wimmer, E., and Dasgupta, A. (1991). Poliovirus proteinase 3C converts an active form of transcription factor IIIC to an inactive form: a mechanism for inhibition of host cell polymerase III transcription by poliovirus. EMBO J. *10*, 2941–2947.

Cortes, Ledesma, F., El Khamisy, S.F., Zuma, M.C., Osborn, K., and Caldecott, K.W. (2009). A human 5′ -tyrosyl DNA phosphodiesterase that repairs topoisomerase-mediated DNA damage. Nature *461*, 674–678. https://doi.org/10.1038/nature08444

Daijogo, S., and Semler, B.L. (2011). Mechanistic intersections between picornavirus translation and RNA replication. Adv. Virus Res. *80*, 1–24. https://doi.org/10.1016/B978-0-12-385987-7.00001-4

Dales, S., Eggers, H.J., Tamm, I., and Palade, G.E. (1965). Electron microscopic study of the formation OF poliovirus. Virology *26*, 379–389.

Das, S., and Dasgupta, A. (1993). Identification of the cleavage site and determinants required for poliovirus 3CPro-catalyzed cleavage of human TATA-binding transcription factor TBP. J. Virol. *67*, 3326–3331.

de Breyne, S., Yu, Y., Unbehaun, A., Pestova, T.V., and Hellen, C.U. (2009). Direct functional interaction of initiation factor eIF4G with type 1 internal ribosomal entry sites. Proc. Natl. Acad. Sci. U.S.A. *106*, 9197–9202. https://doi.org/10.1073/pnas.0900153106

Deddouche, S., Goubau, D., Rehwinkel, J., Chakravarty, P., Begum, S., Maillard, P.V., Borg, A., Matthews, N., Feng, Q., van Kuppeveld, F.J., *et al.* (2014). Identification of an LGP2-associated MDA5 agonist in picornavirus-infected cells. Elife *3*, e01535. https://doi.org/10.7554/eLife.01535

Dempsey, L.A., Hanakahi, L.A., and Maizels, N. (1998). A specific isoform of hnRNP D interacts with DNA in the LR1 heterodimer: canonical RNA binding motifs in a sequence-specific duplex DNA binding protein. J. Biol. Chem. *273*, 29224–29229.

Devaney, M.A., Vakharia, V.N., Lloyd, R.E., Ehrenfeld, E., and Grubman, M.J. (1988). Leader protein of foot-and-mouth disease virus is required for cleavage of the p220 component of the cap-binding protein complex. J. Virol. *62*, 4407–4409.

Dorobantu, C.M., van der Schaar, H.M., Ford, L.A., Strating, J.R., Ulferts, R., Fang, Y., Belov, G., and van Kuppeveld, F.J. (2014). Recruitment of PI4KIIIβ to coxsackievirus B3 replication organelles is independent of ACBD3, GBF1, and Arf1. J. Virol. *88*, 2725–2736. https://doi.org/10.1128/JVI.03650-13

Dougherty, J.D., White, J.P., and Lloyd, R.E. (2011). Poliovirus-mediated disruption of cytoplasmic processing bodies. J. Virol. *85*, 64–75. https://doi.org/10.1128/JVI.01657-10

Drahos, J., and Racaniello, V.R. (2009). Cleavage of IPS-1 in cells infected with human rhinovirus. J. Virol. *83*, 11581–11587. https://doi.org/10.1128/JVI.01490-09

Ernst, V., Levin, D.H., and London, I.M. (1978). Inhibition of protein synthesis initiation by oxidized glutathione: activation of a protein kinase that phosphorylates the alpha subunit of eukaryotic initiation factor 2. Proc. Natl. Acad. Sci. U.S.A. *75*, 4110-4114.

Etchison, D., Milburn, S.C., Edery, I., Sonenberg, N., and Hershey, J.W. (1982). Inhibition of HeLa cell protein synthesis following poliovirus infection correlates with the proteolysis of a 220,000-dalton polypeptide associated with eucaryotic initiation factor 3 and a cap binding protein complex. J. Biol. Chem. *257*, 14806-14810.

Feng, Q., Hato, S.V., Langereis, M.A., Zoll, J., Virgen-Slane, R., Peisley, A., Hur, S., Semler, B.L., van Rij, R.P., and van Kuppeveld, F.J. (2012). MDA5 detects the double-stranded RNA replicative form in picornavirus-infected cells. Cell Rep 2, 1187–1196. https://doi.org/10.1016/j.celrep.2012.10.005

Feng, Q., Langereis, M.A., Lork, M., Nguyen, M., Hato, S.V., Lanke, K., Emdad, L., Bhoopathi, P., Fisher, P.B., Lloyd, R.E., *et al.* (2014). Enterovirus 2Apro targets MDA5 and MAVS in infected cells. J. Virol. *88*, 3369–3378. https://doi.org/10.1128/JVI.02712-13

Fernandez-Munoz, R., and Darnell, J.E. (1976). Structural difference between the 5′ termini of viral and cellular mRNA in poliovirus-infected cells: possible basis for the inhibition of host protein synthesis. J. Virol. *18*, 719-726.

Fitzgerald, K.D., and Semler, B.L. (2009). Bridging IRES elements in mRNAs to the eukaryotic translation apparatus. Biochim. Biophys. Acta *1789*, 518–528. https://doi.org/10.1016/j.bbagrm.2009.07.004

Fitzgerald, K.D., and Semler, B.L. (2011). Re-localization of cellular protein SRp20 during poliovirus infection: bridging a viral IRES to the host cell translation apparatus. PLOS Pathog. *7*, e1002127. https://doi.org/10.1371/journal.ppat.1002127

Fitzgerald, K.D., and Semler, B.L. (2013). Poliovirus infection induces the co-localization of cellular protein SRp20 with TIA-1, a cytoplasmic stress granule protein. Virus Res. *176*, 223–231. https://doi.org/10.1016/j.virusres.2013.06.012

Fitzgerald, K.D., Chase, A.J., Cathcart, A.L., Tran, G.P., and Semler, B.L. (2013). Viral proteinase requirements for the nucleocytoplasmic relocalization of cellular splicing factor SRp20 during picornavirus infections. J. Virol. *87*, 2390-2400. https://doi.org/10.1128/JVI.02396-12

Flanegan, J.B., and Baltimore, D. (1977). Poliovirus-specific primer-dependent RNA polymerase able to copy poly(A). Proc. Natl. Acad. Sci. U.S.A. *74*, 3677–3680.

Flanegan, J.B., Petterson, R.F., Ambros, V., Hewlett, N.J., and Baltimore, D. (1977). Covalent linkage of a protein to a defined nucleotide sequence at the 5′-terminus of virion and replicative intermediate RNAs of poliovirus. Proc. Natl. Acad. Sci. U.S.A. *74*, 961–965.

Florez, P.M., Sessions, O.M., Wagner, E.J., Gromeier, M., and Garcia-Blanco, M.A. (2005). The polypyrimidine tract binding protein is required for efficient picornavirus gene expression and propagation. J. Virol. *79*, 6172–6179.

Fung, G., Ng, C.S., Zhang, J., Shi, J., Wong, J., Piesik, P., Han, L., Chu, F., Jagdeo, J., Jan, E., Fujita, T., and Luo, H.(2013). Production of a dominant-negative fragment due to G3BP1 cleavage contributes to the disruption of mitochondria-associated protective stress granules during CVB3 infection. PLOS ONE *8*, e79546. https://doi.org/10.1371/journal.pone.0079546

Gamarnik, A.V., and Andino, R. (1997). Two functional complexes formed by KH domain containing proteins with the 5′ noncoding region of poliovirus RNA. RNA 3, 882–892.

Gamarnik, A.V., and Andino, R. (1998). Switch from translation to RNA replication in a positive-stranded RNA virus. Genes Dev. *12*, 2293–2304.

Gavala, M.L., Bertics, P.J., and Gern, J.E. (2011). Rhinoviruses, allergic inflammation, and asthma. Immunol. Rev. *242*, 69–90. https://doi.org/10.1111/j.1600-065X.2011.01031.x

Golini, F., Semler, B.L., Dorner, A.J., and Wimmer, E. (1980). Protein-linked RNA of poliovirus is competent to form an initiation complex of translation in vitro. Nature *287*, 600–603.

Görlich, D., and Kutay, U. (1999). Transport between the cell nucleus and the cytoplasm. Annu. Rev. Cell Dev. Biol. *15*, 607–660. https://doi.org/10.1146/annurev.cellbio.15.1.607

Gottlieb, E., and Steitz, J.A. (1989). The RNA binding protein La influences both the accuracy and the efficiency of RNA polymerase III transcription in vitro. EMBO J. *8*, 841–850.

Greninger, A.L., Knudsen, G.M., Betegon, M., Burlingame, A.L., and Derisi, J.L. (2012). The 3A protein from multiple picornaviruses utilizes the golgi adaptor protein ACBD3 to recruit PI4KIIIβ. J. Virol. *86*, 3605–3616. https://doi.org/10.1128/JVI.06778-11

Gustin, K.E. (2003). Inhibition of nucleo-cytoplasmic trafficking by RNA viruses: targeting the nuclear pore complex. Virus Res. *95*, 35–44.

Gustin, K.E., and Sarnow, P. (2001). Effects of poliovirus infection on nucleo-cytoplasmic trafficking and nuclear pore complex composition. EMBO J. *20*, 240–249. https://doi.org/10.1093/emboj/20.1.240

Hewlett, M.J., Rose, J.K., and Baltimore, D. (1976). 5′-terminal structure of poliovirus polyribosomal RNA is pUp. Proc. Natl. Acad. Sci. U.S.A. *73*, 327–330.

Hornung, V., Ellegast, J., Kim, S., Brzózka, K., Jung, A., Kato, H., Poeck, H., Akira, S., Conzelmann, K.K., Schlee, M., *et al.* (2006). 5′-Triphosphate RNA is the ligand for RIG-I. Science *314*, 994–997.

Hsu, N.Y., Ilnytska, O., Belov, G., Santiana, M., Chen, Y.H., Takvorian, P.M., Pau, C., van der Schaar, H., Kaushik-Basu, N., Balla, T., *et al.* (2010). Viral reorganization of the secretory pathway generates distinct organelles for RNA replication. Cell *141*, 799–811. https://doi.org/10.1016/j.cell.2010.03.050

Huang, P.N., Lin, J.Y., Locker, N., Kung, Y.A., Hung, C.T., Lin, J.Y., Huang, H.I., Li, M.L., and Shih, S.R. (2011). Far upstream element binding protein 1 binds the internal ribosomal entry site of enterovirus 71 and enhances viral translation and viral growth. Nucleic Acids Res. *39*, 9633-9648. https://doi.org/10.1093/nar/gkr682

Hunt, S.L., Hsuan, J.J., Totty, N., and Jackson, R.J. (1999). unr, a cellular cytoplasmic RNA-binding protein with five cold-shock domains, is required for internal initiation of translation of human rhinovirus RNA. Genes Dev. *13*, 437–448.

Izumi, R.E., Valdez, B., Banerjee, R., Srivastava, M., and Dasgupta, A. (2001). Nucleolin stimulates viral internal ribosome entry site-mediated translation. Virus Res. *76*, 17–29.

Jackson, W.T., Giddings, T.H., Taylor, M.P., Mulinyawe, S., Rabinovitch, M., Kopito, R.R., and Kirkegaard, K. (2005). Subversion of cellular autophagosomal machinery by RNA viruses. PLOS Biol. *3*, e156.

Joachims, M., Van Breugel, P.C., and Lloyd, R.E. (1999). Cleavage of poly(A)-binding protein by enterovirus proteases concurrent with inhibition of translation in vitro. J. Virol. *73*, 718–727.

Kafasla, P., Morgner, N., Robinson, C.V., and Jackson, R.J. (2010). Polypyrimidine tract-binding protein stimulates the poliovirus IRES by modulating eIF4G binding. EMBO J. *29*, 3710–3722. https://doi.org/10.1038/emboj.2010.231

Kajita, Y., Nakayama, J., Aizawa, M., and Ishikawa, F. (1995). The UUAG-specific RNA binding protein, heterogeneous nuclear ribonucleoprotein D0. Common modular structure and binding properties of the 2xRBD-Gly family. J. Biol. Chem. *270*, 22167-22175.

Kassavetis, G.A., Braun, B.R., Nguyen, L.H., and Geiduschek, E.P. (1990). S. cerevisiae TFIIIB is the transcription initiation factor proper of RNA polymerase III, while TFIIIA and TFIIIC are assembly factors. Cell *60*, 235–245.

Kato, H., Takeuchi, O., Sato, S., Yoneyama, M., Yamamoto, M., Matsui, K., Uematsu, S., Jung, A., Kawai, T., Ishii, K.J., *et al.* (2006). Differential roles of MDA5 and RIG-I helicases in the recognition of RNA viruses. Nature *441*, 101–105.

Kedersha, N.L., Gupta, M., Li, W., Miller, I., and Anderson, P. (1999). RNA-binding proteins TIA-1 and TIAR link the phosphorylation of eIF-2 alpha to the assembly of mammalian stress granules. J. Cell Biol. *147*, 1431–1442.

Kemball, C.C., Alirezaei, M., Flynn, C.T., Wood, M.R., Harkins, S., Kiosses, W.B., and Whitton, J.L. (2010). Coxsackievirus infection induces autophagy-like vesicles and megaphagosomes in pancreatic acinar cells in vivo. J. Virol. *84*, 12110–12124. https://doi.org/10.1128/JVI.01417-10

Kenan, D.J., Query, C.C., and Keene, J.D. (1991). RNA recognition: towards identifying determinants of specificity. Trends Biochem. Sci. *16*, 214–220.

Kerekatte, V., Keiper, B.D., Badorff, C., Cai, A., Knowlton, K.U., and Rhoads, R.E. (1999). Cleavage of Poly(A)-binding protein by coxsackievirus 2A protease in vitro and in vivo: another mechanism for host protein synthesis shutoff? J. Virol. *73*, 709–717.

Khetsuriani, N., Lamonte-Fowlkes, A., Oberst, S., and Pallansch, M.A. (2006). Enterovirus surveillance – United States, 1970-2005. MMWR Surveill Summ *55*, 1–20.

Kiledjian, M., DeMaria, C.T., Brewer, G., and Novick, K. (1997). Identification of AUF1 (heterogeneous nuclear ribonucleoprotein D) as a component of the alpha-globin mRNA stability complex. Mol. Cell. Biol. *17*, 4870–4876.

Kim, J.H., Hahm, B., Kim, Y.K., Choi, M., and Jang, S.K. (2000). Protein-protein interaction among hnRNPs shuttling between nucleus and cytoplasm. J. Mol. Biol. *298*, 395–405. https://doi.org/10.1006/jmbi.2000.3687

Kitamura, N., Semler, B.L., Rothberg, P.G., Larsen, G.R., Adler, C.J., Dorner, A.J., Emini, E.A., Hanecak, R., Lee, J.J., van der Werf, S., *et al.* (1981). Primary structure, gene organization and polypeptide expression of poliovirus RNA. Nature *291*, 547–553.

Klein, K.A., and Jackson, W.T. (2011). Human rhinovirus 2 induces the autophagic pathway and replicates more efficiently in autophagic cells. J. Virol. *85*, 9651–9654. https://doi.org/10.1128/JVI.00316-11

Klumperman, J. (2000). Transport between ER and Golgi. Curr. Opin. Cell Biol. *12*, 445–449.

Kozlov, G., De Crescenzo, G., Lim, N.S., Siddiqui, N., Fantus, D., Kahvejian, A., Trempe, J.F., Elias, D., Ekiel, I., Sonenberg, N., *et al.* (2004). Structural basis of ligand recognition by PABC, a highly specific peptide-binding domain found in poly(A)-binding protein and a HECT ubiquitin ligase. EMBO J. *23*, 272-281.

Kozlov, G., Trempe, J.F., Khaleghpour, K., Kahvejian, A., Ekiel, I., and Gehring, K. (2001). Structure and function of the C-terminal PABC domain of human poly(A)-binding protein. Proc. Natl. Acad. Sci. U.S.A. *98*, 4409–4413. https://doi.org/10.1073/pnas.071024998

Kräusslich, H.G., Nicklin, M.J., Toyoda, H., Etchison, D., and Wimmer, E. (1987). Poliovirus proteinase 2A induces cleavage of eucaryotic initiation factor 4F polypeptide p220. J. Virol. *61*, 2711–2718.

Kuo, R.L., Kao, L.T., Lin, S.J., Wang, R.Y., and Shih, S.R. (2013). MDA5 plays a crucial role in enterovirus 71 RNA-mediated IRF3 activation. PLOS ONE *8*, e63431. https://doi.org/10.1371/journal.pone.0063431

Kuyumcu-Martinez, N.M., Joachims, M., and Lloyd, R.E. (2002). Efficient cleavage of ribosome-associated poly(A)-binding protein by enterovirus 3C protease. J. Virol. *76*, 2062–2074.

Lamphear, B.J., Kirchweger, R., Skern, T., and Rhoads, R.E. (1995). Mapping of functional domains in eukaryotic protein synthesis initiation factor 4G (eIF4G) with picornaviral proteases. Implications for cap-dependent and cap-independent translational initiation. J. Biol. Chem. *270*, 21975-21983.

Langereis, M.A., Feng, Q., Nelissen, F.H., Virgen-Slane, R., van der Heden van Noort, G.J., Maciejewski, S., Filippov, D.V., Semler, B.L., van Delft, F.L., and van Kuppeveld, F.J. (2014). Modification of picornavirus genomic RNA using 'click' chemistry shows that unlinking of the VPg peptide is dispensable for translation and replication of the incoming viral RNA. Nucleic Acids Res *42*, 2473-2482. https://doi.org/10.1093/nar/gkt1162

Lee, Y.F., Nomoto, A., Detjen, B.M., and Wimmer, E. (1977). A protein covalently linked to poliovirus genome RNA. Proc. Natl. Acad. Sci. U.S.A. *74*, 59–63.

Lei, X., Liu, X., Ma, Y., Sun, Z., Yang, Y., Jin, Q., He, B., and Wang, J. (2010). The 3C protein of enterovirus 71 inhibits retinoid acid-inducible gene I-mediated interferon regulatory factor 3 activation and type I interferon responses. J. Virol. *84*, 8051–8061. https://doi.org/10.1128/JVI.02491-09

Lei, X., Xiao, X., Xue, Q., Jin, Q., He, B., and Wang, J. (2013). Cleavage of interferon regulatory factor 7 by enterovirus 71 3C suppresses cellular responses. J. Virol. *87*, 1690–1698. https://doi.org/10.1128/JVI.01855-12

Lenarcic, E.M., Landry, D.M., Greco, T.M., Cristea, I.M., and Thompson, S.R. (2013). Thiouracil cross-linking mass spectrometry: a cell-based method to identify host factors involved in viral amplification. J. Virol. *87*, 8697–8712. https://doi.org/10.1128/JVI.00950-13

Liebig, H.D., Ziegler, E., Yan, R., Hartmuth, K., Klump, H., Kowalski, H., Blaas, D., Sommergruber, W., Frasel, L., Lamphear, B., Rhoads, R., Kuechler, E., and Skern, T. (1993). Purification of two picornaviral 2A proteinases: interaction with eIF-4 gamma and influence on in vitro translation. Biochemistry *32*, 7581-7588.

Lin, J.Y., Li, M.L., and Shih, S.R. (2009). Far upstream element binding protein 2 interacts with enterovirus 71 internal ribosomal entry site and negatively regulates viral translation. Nucleic Acids Res. *37*, 47-59. https://doi.org/10.1093/nar/gkn901

Lin, J.Y., Li, M.L., Huang, P.N., Chien, K.Y., Horng, J.T., and Shih, S.R. (2008). Heterogeneous nuclear ribonuclear protein K interacts with the enterovirus 71 5′ untranslated region and participates in virus replication. J. Gen. Virol. *89*, 2540-2549. https://doi.org/10.1099/vir.0.2008/003673-0

Lloyd, R.E. (2006). Translational control by viral proteinases. Virus Res. *119*, 76–88.

Lu, J.Y., Bergman, N., Sadri, N., and Schneider, R.J. (2006). Assembly of AUF1 with eIF4G-poly(A) binding protein complex suggests a translation function in AU-rich mRNA decay. RNA *12*, 883–893.

Maciejewski, S., Nguyen, J.H., Gómez-Herreros, F., Cortés-Ledesma, F., Caldecott, K.W., and Semler, B.L. (2015). Divergent requirement for a DNA repair enzyme during Enterovirus infections. mBio 7(*1*):e01931-15. https://doi.org/10.1128/mBio.01931-15

Mazroui, R., Sukarieh, R., Bordeleau, M.E., Kaufman, R.J., Northcote, P., Tanaka, J., Gallouzi, I., and Pelletier, J. (2006). Inhibition of ribosome recruitment induces stress granule formation independently of eukaryotic initiation factor 2alpha phosphorylation. Mol. Biol. Cell. *17*, 4212-4219.

McBride, A.E., Schlegel, A., and Kirkegaard, K. (1996). Human protein Sam68 relocalization and interaction with poliovirus RNA polymerase in infected cells. Proc. Natl. Acad. Sci. U.S.A. *93*, 2296–2301.

Meerovitch, K., Svitkin, Y.V., Lee, H.S., Lejbkowicz, F., Kenan, D.J., Chan, E.K., Agol, V.I., Keene, J.D., and Sonenberg, N. (1993). La autoantigen enhances and corrects aberrant translation of poliovirus RNA in reticulocyte lysate. J. Virol. *67*, 3798–3807.

Michael, W.M., Eder, P.S., and Dreyfuss, G. (1997). The K nuclear shuttling domain: a novel signal for nuclear import and nuclear export in the hnRNP K protein. EMBO J. *16*, 3587–3598. https://doi.org/10.1093/emboj/16.12.3587

Midgley, C.M., Jackson, M.A., Selvarangan, R., Turabelidze, G., Obringer, E., Johnson, D., Giles, B.L., Patel, A., Echols, F., Oberste, M.S., *et al.* (2014). Severe respiratory illness associated with enterovirus D68 - Missouri and Illinois, 2014. MMWR Morb. Mortal. Wkly. Rep. *63*, 798–799.

Mukherjee, A., Morosky, S.A., Delorme-Axford, E., Dybdahl-Sissoko, N., Oberste, M.S., Wang, T., and Coyne, C.B. (2011). The coxsackievirus B 3C protease cleaves MAVS and TRIF to attenuate host type I interferon and apoptotic signaling. PLOS Pathog. *7*, e1001311. https://doi.org/10.1371/journal.ppat.1001311

Mulligan, G.J., Guo, W., Wormsley, S., and Helfman, D.M. (1992). Polypyrimidine tract binding protein interacts with sequences involved in alternative splicing of beta-tropomyosin pre-mRNA. J. Biol. Chem. *267*, 25480–25487.

Nicklin, M.J.H., Toyoda, H., Murray, M.G., and Wimmer, E. (1986). Proteolytic processing in the replication of polio and related viruses. Nat. Biotechnol. *4*, 36-42. https://doi.org/10.1038/nbt0186-33

Nomoto, A., Kitamura, N., Golini, F., and Wimmer, E. (1977). The 5′ -terminal structures of poliovirion RNA and poliovirus mRNA differ only in the genome-linked protein VPg. Proc. Natl. Acad. Sci. U.S.A. *74*, 5345–5349.

Nomoto, A., Lee, Y.F., and Wimmer, E. (1976). The 5′ end of poliovirus mRNA is not capped with m7G(5′)ppp(5′)Np. Proc. Natl. Acad. Sci. U.S.A. *73*, 375–380.

Oshiumi, H., Okamoto, M., Fujii, K., Kawanishi, T., Matsumoto, M., Koike, S., and Seya, T. (2011). The TLR3/TICAM-1 pathway is mandatory for innate immune responses to poliovirus infection. J. Immunol. *187*, 5320–5327. https://doi.org/10.4049/jimmunol.1101503

Park, N., Katikaneni, P., Skern, T., and Gustin, K.E. (2008). Differential targeting of nuclear pore complex proteins in poliovirus-infected cells. J. Virol. *82*, 1647–1655.

Park, N., Skern, T., and Gustin, K.E. (2010). Specific cleavage of the nuclear pore complex protein Nup62 by a viral protease. J. Biol. Chem. *285*, 28796–28805. https://doi.org/10.1074/jbc.M110.143404

Parsley, T.B., Towner, J.S., Blyn, L.B., Ehrenfeld, E., and Semler, B.L. (1997). Poly (rC) binding protein 2 forms a ternary complex with the 5′ -terminal sequences of poliovirus RNA and the viral 3CD proteinase. RNA 3, 1124–1134.

Paul, A.V., van Boom, J.H., Filippov, D., and Wimmer, E. (1998). Protein-primed RNA synthesis by purified poliovirus RNA polymerase. Nature *393*, 280–284. https://doi.org/10.1038/30529

Peisley, A., Wu, B., Yao, H., Walz, T., and Hur, S. (2013). RIG-I forms signaling-competent filaments in an ATP-dependent, ubiquitin-independent manner. Mol. Cell *51*, 573–583. https://doi.org/10.1016/j.molcel.2013.07.024

Perera, R., Daijogo, S., Walter, B.L., Nguyen, J.H., and Semler, B.L. (2007). Cellular protein modification by poliovirus: the two faces of poly(rC)-binding protein. J. Virol. *81*, 8919–8932.

Piotrowska, J., Hansen, S.J., Park, N., Jamka, K., Sarnow, P., and Gustin, K.E. (2010). Stable formation of compositionally unique stress granules in virus-infected cells. J. Virol. *84*, 3654–3665. https://doi.org/10.1128/JVI.01320-09

Pollard, V.W., Michael, W.M., Nakielny, S., Siomi, M.C., Wang, F., and Dreyfuss, G. (1996). A novel receptor-mediated nuclear protein import pathway. Cell *86*, 985–994.

Ray, P.S., and Das, S. (2002). La autoantigen is required for the internal ribosome entry site-mediated translation of Coxsackievirus B3 RNA. Nucleic Acids Res. *30*, 4500–4508.

Rebsamen, M., Meylan, E., Curran, J., and Tschopp, J. (2008). The antiviral adaptor proteins Cardif and Trif are processed and inactivated by caspases. Cell Death Differ. *15*, 1804–1811. https://doi.org/10.1038/cdd.2008.119

Reikine, S., Nguyen, J.B., and Modis, Y. (2014). Pattern Recognition and Signaling Mechanisms of RIG-I and MDA5. Front Immunol 5, 342. https://doi.org/10.3389/fimmu.2014.00342

Richards, A.L., Soares-Martins, J.A., Riddell, G.T., and Jackson, W.T. (2014). Generation of unique poliovirus RNA replication organelles. MBio *5*, e00833–13. https://doi.org/10.1128/mBio.00833-13

Rothberg, P.G., Harris, T.J., Nomoto, A., and Wimmer, E. (1978). O4-(5 -uridylyl)tyrosine is the bond between the genome-linked protein and the RNA of poliovirus. Proc. Natl. Acad. Sci. U.S.A. *75*, 4868–4872.

Rozovics, J.M., Chase, A.J., Cathcart, A.L., Chou, W., Gershon, P.D., Palusa, S., Wilusz, J., and Semler, B.L. (2012). Picornavirus modification of a host mRNA decay protein. MBio *3*, e00431–12. https://doi.org/10.1128/mBio.00431-12

Rozovics, J.M., Virgen-Slane, R., and Semler, B.L. (2011). Engineered picornavirus VPg-RNA substrates: analysis of a tyrosyl-RNA phosphodiesterase activity. PLOS ONE *6*, e16559. https://doi.org/10.1371/journal.pone.0016559

Rust, R.C., Landmann, L., Gosert, R., Tang, B.L., Hong, W., Hauri, H.P., Egger, D., and Bienz, K. (2001). Cellular COPII proteins are involved in production of the vesicles that form the poliovirus replication complex. J. Virol. *75*, 9808–9818. https://doi.org/10.1128/JVI.75.20.9808-9818.2001

Schlegel, A., Giddings, T.H., Ladinsky, M.S., and Kirkegaard, K. (1996). Cellular origin and ultrastructure of membranes induced during poliovirus infection. J. Virol. *70*, 6576–6588.

Shen, H., Kan, J.L., and Green, M.R. (2004). Arginine-serine-rich domains bound at splicing enhancers contact the branchpoint to promote prespliceosome assembly. Mol. Cell *13*, 367–376.

Shih, S.R., Stollar, V., and Li, M.L. (2011). Host factors in enterovirus 71 replication. J. Virol. *85*, 9658–9666. https://doi.org/10.1128/JVI.05063-11

Shiroki, K., Isoyama, T., Kuge, S., Ishii, T., Ohmi, S., Hata, S., Suzuki, K., Takasaki, Y., and Nomoto, A. (1999). Intracellular redistribution of truncated La protein produced by poliovirus 3Cpro-mediated cleavage. J. Virol. *73*, 2193–2200.

Smith, R.W., Blee, T.K., and Gray, N.K. (2014). Poly(A)-binding proteins are required for diverse biological processes in metazoans. Biochem. Soc. Trans. *42*, 1229–1237. https://doi.org/10.1042/BST20140111

Springer, S., Spang, A., and Schekman, R. (1999). A primer on vesicle budding. Cell *97*, 145–148.

Stewart, M., Baker, R.P., Bayliss, R., Clayton, L., Grant, R.P., Littlewood, T., and Matsuura, Y. (2001). Molecular mechanism of translocation through nuclear pore complexes during nuclear protein import. FEBS Lett. *498*, 145-149.

Suhy, D.A., Giddings, T.H., and Kirkegaard, K. (2000). Remodeling the endoplasmic reticulum by poliovirus infection and by individual viral proteins: an autophagy-like origin for virus-induced vesicles. J. Virol. *74*, 8953–8965.

Svitkin, Y.V., Meerovitch, K., Lee, H.S., Dholakia, J.N., Kenan, D.J., Agol, V.I., and Sonenberg, N. (1994). Internal translation initiation on poliovirus RNA: further characterization of La function in poliovirus translation in vitro. J. Virol. *68*, 1544–1550.

Sweeney, T.R., Abaeva, I.S., Pestova, T.V., and Hellen, C.U. (2014). The mechanism of translation initiation on Type 1 picornavirus IRESs. EMBO J. *33*, 76–92. https://doi.org/10.1002/embj.201386124

Tan, E.M. (1989). Antinuclear antibodies: diagnostic markers for autoimmune diseases and probes for cell biology. Adv. Immunol. *44*, 93–151.

Taylor, M.P., and Kirkegaard, K. (2007). Modification of cellular autophagy protein LC3 by poliovirus. J. Virol. *81*, 12543–12553.

Tokarz, R., Firth, C., Madhi, S.A., Howie, S.R., Wu, W., Sall, A.A., Haq, S., Briese, T., and Lipkin, W.I. (2012). Worldwide emergence of multiple clades of enterovirus 68. J. Gen. Virol. *93*, 1952-1958. https://doi.org/10.1099/vir.0.043935-0

Tourrière, H., Chebli, K., Zekri, L., Courselaud, B., Blanchard, J.M., Bertrand, E., and Tazi, J. (2003). The RasGAP-associated endoribonuclease G3BP assembles stress granules. J. Cell Biol. *160*, 823–831. https://doi.org/10.1083/jcb.200212128

Toyoda, H., Nicklin, M.J., Murray, M.G., Anderson, C.W., Dunn, J.J., Studier, F.W., and Wimmer, E. (1986). A second virus-encoded proteinase involved in proteolytic processing of poliovirus polyprotein. Cell *45*, 761–770.

Trahey, M., Oh, H.S., Cameron, C.E., and Hay, J.C. (2012). Poliovirus infection transiently increases COPII vesicle budding. J. Virol. *86*, 9675–9682. https://doi.org/10.1128/JVI.01159-12

Triantafilou, K., Vakakis, E., Kar, S., Richer, E., Evans, G.L., and Triantafilou, M. (2012). Visualisation of direct interaction of MDA5 and the dsRNA replicative intermediate form of positive strand RNA viruses. J. Cell. Sci. *125*, 4761-4769. https://doi.org/10.1242/jcs.103887

Virgen-Slane, R., Rozovics, J.M., Fitzgerald, K.D., Ngo, T., Chou, W., van der Heden van Noort, G.J., Filippov, D.V., Gershon, P.D., and Semler, B.L. (2012). An RNA virus hijacks an incognito function of a DNA repair enzyme. Proc. Natl. Acad. Sci. U.S.A. *109*, 14634–14639. https://doi.org/10.1073/pnas.1208096109

Vogt, D.A., and Andino, R. (2010). An RNA element at the 5′-end of the poliovirus genome functions as a general promoter for RNA synthesis. PLOS Pathog. *6*, e1000936. https://doi.org/10.1371/journal.ppat.1000936

Waggoner, S., and Sarnow, P. (1998). Viral ribonucleoprotein complex formation and nucleolar-cytoplasmic relocalization of nucleolin in poliovirus-infected cells. J. Virol. 72, 6699-6709.

Wang, B., Xi, X., Lei, X., Zhang, X., Cui, S., Wang, J., Jin, Q., and Zhao, Z. (2013). Enterovirus 71 protease 2Apro targets MAVS to inhibit anti-viral type I interferon responses. PLOS Pathog. *9*, e1003231. https://doi.org/10.1371/journal.ppat.1003231

Wang, J.P., Cerny, A., Asher, D.R., Kurt-Jones, E.A., Bronson, R.T., and Finberg, R.W. (2010). MDA5 and MAVS mediate type I interferon responses to coxsackie B virus. J. Virol. *84*, 254–260. https://doi.org/10.1128/JVI.00631-09

Watters, K., and Palmenberg, A.C. (2011). Differential processing of nuclear pore complex proteins by rhinovirus 2A proteases from different species and serotypes. J. Virol. *85*, 10874–10883. https://doi.org/10.1128/JVI.00718-11

Weidman, M.K., Yalamanchili, P., Ng, B., Tsai, W., and Dasgupta, A. (2001). Poliovirus 3C protease-mediated degradation of transcriptional activator p53 requires a cellular activity. Virology *291*, 260–271. https://doi.org/10.1006/viro.2001.1215

Wells, S.E., Hillner, P.E., Vale, R.D., and Sachs, A.B. (1998). Circularization of mRNA by eukaryotic translation initiation factors. Mol. Cell *2*, 135–140.

White, J.P., and Lloyd, R.E. (2011). Poliovirus unlinks TIA1 aggregation and mRNA stress granule formation. J. Virol. *85*, 12442–12454. https://doi.org/10.1128/JVI.05888-11

White, J.P., Cardenas, A.M., Marissen, W.E., and Lloyd, R.E. (2007). Inhibition of cytoplasmic mRNA stress granule formation by a viral proteinase. Cell Host Microbe *2*, 295–305.

Wimmer, E., Hellen, C.U., and Cao, X. (1993). Genetics of poliovirus. Annu. Rev. Genet. *27*, 353–436. https://doi.org/10.1146/annurev.ge.27.120193.002033

Wong, J., Si, X., Angeles, A., Zhang, J., Shi, J., Fung, G., Jagdeo, J., Wang, T., Zhong, Z., Jan, E., Luo, H. (2013). Cytoplasmic redistribution and cleavage of AUF1 during coxsackievirus infection enhance the stability of its viral genome. FASEB J. *27*, 2777-2787. https://doi.org/10.1096/fj.12-226498

Yalamanchili, P., Banerjee, R., and Dasgupta, A. (1997a). Poliovirus-encoded protease 2APro cleaves the TATA-binding protein but does not inhibit host cell RNA polymerase II transcription in vitro. J. Virol. *71*, 6881–6886.

Yalamanchili, P., Datta, U., and Dasgupta, A. (1997b). Inhibition of host cell transcription by poliovirus: cleavage of transcription factor CREB by poliovirus-encoded protease 3Cpro. J. Virol. *71*, 1220–1226.

Yalamanchili, P., Harris, K., Wimmer, E., and Dasgupta, A. (1996). Inhibition of basal transcription by poliovirus: a virus- encoded protease (3Cpro) inhibits formation of TBP-TATA box complex in vitro. J. Virol. *70*, 2922–2929.

Yogo, Y., and Wimmer, E. (1972). Polyadenylic acid at the 3'-terminus of poliovirus RNA. Proc. Natl. Acad. Sci. U.S.A. *69*, 1877–1882.

Zhang, W., Wagner, B.J., Ehrenman, K., Schaefer, A.W., DeMaria, C.T., Crater, D., DeHaven, K., Long, L., and Brewer, G. (1993). Purification, characterization, and cDNA cloning of an AU-rich element RNA-binding protein, AUF1. Mol. Cell. Biol. *13*, 7652–7665.

The Omics of Rhinoviruses

Ann C. Palmenberg

4

Institute for Molecular Virology, R.M. Bock Laboratories, University of Wisconsin–Madison, Madison, WI, USA.

Correspondence: acpalmen@wisc.edu

https://doi.org/10.21775/9781910190739.04

Abstract

The human rhinoviruses comprise three species, RV-A, RV-B and RV-C, in the *Enterovirus* genus of *Picornaviridae.* Prior to the current 'Omics Era' viruses were typically binned by dominant disease phenotypes after a simple parsing of their RNA or DNA genotypes. But as the 'omics' has piled up, and more and more sequences have emerged, the similarities in genome organization, receptor use, aetiology and structures has become ever more apparent. This chapter summarizes key aspects of RV taxonomy and the physical and biochemical parameters by which these viruses are currently studied.

The Rhinoviruses

The human rhinoviruses comprise three species, RV-A, RV-B and RV-C, in the *Enterovirus* genus of *Picornaviridae.* That taxonomic statement has probably opened at least ten review articles in recent years. In truth though, it is a relatively new arrangement. The RV-C are only recent additions to the Enteroviruses. Moreover, there was a time when the RV-A and RV-B were actually accorded their own genus. Prior to the current 'Omics Era' viruses were typically binned by dominant disease phenotypes after a simple parsing of their RNA or DNA genotypes. To justify research funding priorities, the respective territories were tenaciously defended by advocates working in those fields. RV 'common cold' agents were believed to warrant an independent classification that distinguished them from the more severe (and more important) enteric diseases caused by polio-, coxsackie- or enteric cytopathic human orphan (ECHO) viruses. Therefore, it surprised many research groups when in 1984–85, the first sequence of RV-B14 along with its virion crystal structure, showed remarkable similarity to the parallel polio datasets that were emerging at the same time (Rossmann *et al.*, 1985; Stanway *et al.*, 1984). At first the commonalities were downplayed with the expectation that maybe all picornaviruses would prove similar. That idea was disproven when Mengo virus (a *Cardiovirus*), foot-and-mouth disease virus (FMDV, an *Aphthovirus*) and hepatitis A (an *Hepatovirus*) were resolved in sequence and structure, and shown to be really quite dissimilar, as befitted their separate genera status (Luo *et al.*, 1987, 1988). Picornavirus meetings about this time occasionally dissolved into roaring debates about the pros and cons of assimilating the polios and rhinos into a single supergenus ('*Renterovirus*') that might accord lower status to both factions. It was argued that clinicians would be confused by any new system not based on disease diagnosis. But as the 'omics' piled up, and more and more sequences emerged, the similarities in genome organization, receptor use, aetiology and structures became ever more apparent. In 2009, common sense prevailed, and on the consensus advice of the Picornavirus Study Group, the International Committee on the Taxonomy of Viruses (ICTV) ratified the RV-A and RV-B as two distinct species of *Enterovirus*, retaining '*rhinovirus*' as their naming convention (Adams *et al.*, 2013; King *et al.* 2012).

Just a few years prior to this, the first outbreak of severe acute respiratory syndrome virus (SARS) coronavirus in southern China rocked the virology world. The problem was not the severity of SARS itself, as the outbreak was contained fairly quickly. Instead, regional healthcare workers, trying to avoid embarrassment for their communities, did not report or act on the outbreak until it became global. In response to this lesson, several Asian countries astutely implemented new sweeping clinical surveillance programmes that could catch any re-emergent SARS-related outbreak, if it were to occur (Smith, 2006). The easy application of 'omics' techniques such as polymerase chain reaction (PCR) and gene chips, even in rural clinics, made it possible to quickly and inexpensively identify culpable viruses in patients hospitalized with virtually any respiratory symptoms. The surveillance net quickly resulted in a huge catch of newly identified isolates for influenza, respiratory syncytial virus (RSV), *coronaviruses* and, of course, RV-A and RV-B. The 99 known serotypes were frequently snagged, adding considerable depth to the sequence database. But in addition, a series of 'RV-like' sequences were fished out, initially from clinics in Hong Kong (Lau *et al.*, 2007), and then confirmed world-wide (Lee *et al.*, 2007).

Knowing what to look for brought many subsequent observations, deepening the datasets. In 2006, the first full genome determinations confirmed that these unusual rhinos were subtly but definitively different from the RV-A and RV-B and all other Enteroviruses (Dominguez *et al.*, 2008; Lau *et al.*, 2007). Classified eventually as the RV-C, viruses in this species were among the first to receive official taxa status based solely on sequencing. Unlike the RV-A and RV-B, the RV-C do not grow in standard tissue culture (Bochkov *et al.*, 2011). Yet the RV-C can readily infect both the upper and lower airways, frequently causing pneumonia-like symptoms. These are not new viruses. Most have probably been continuously circulating for thousands of years, just like the RV-A and RV-B. However, before 'omics' techniques were available, all these isolates were simply missed by growth-dependent collection techniques. It was simply assumed, 'no plaque, no virus.' The lack of previous detection (or classification) was because cadherin-related family member 3 (CDHR3), now recognized as the cell receptor for the RV-C, is very different from the receptors used by the RV-A and RV-B [i.e. intercellular adhesion molecule 1 (ICAM-1) and low-density lipoprotein receptor (LDLR)], and this particular protein is apparently down-regulated whenever primary tissue samples are converted to undifferentiated monolayers (Bochkov *et al.*, 2011, 2015). The very recent identification of CDHR3 as the RV-C receptor relied on canonical 'omics' techniques, including expression differentials on gene chips, molecular protein modelling (see below) and comparative sequence analysis. To come full circle with disease phenotypes though, it is now recognized the RV-C have special clinical relevance since many of these strains are associated with severe, hospitalization-category infections in young children, especially those with asthma (Bochkov *et al.*, 2011; Lau *et al.*, 2009). A relatively rare, but dominant, asthma-related allele of the CDHR3 gene apparently causes greater RV-C receptor display on the pulmonary cell surfaces of these children, making them making them 5–10 times more susceptible to devastating virus-triggered asthma episodes (Bochkov *et al.*, 2015; Bonnelykke *et al.*, 2014).

How RV taxa are defined

For the classic RV-A and RV-B, 'genotype' was initially synonymous with 'serotype' because the original designations relied on an ever-growing panel of rabbit-raised antibodies. Neutralization (plaque reduction) or lack thereof, within that panel, placed an isolate into the existing continuum, or added the new serotype to the growing panel (Hamparian *et al.*, 1987; Kapikian, 1967). Given that polio only ever showed three serotypes, as detected by similar methods, after the identification of 100+ rhinoviruses, enthusiasm for this categorization method understandably waned. Thereafter, rhino typing was perhaps not as vigorously maintained as it should have been. The advent of 'omics' in recent years rejuvenated and smoothed many of the wrinkles inadvertently introduced by that system. There are indeed a great many rhinoviruses, perhaps exceeding the genotypes in all other *Enterovirus* species combined (Fig. 4.1). Completion of full genome sequences of at least one isolate for the classic 99, highlighted hidden identities within or between clades that could not have been expected by serotyping alone (Palmenberg *et al.*, 2009). That, and the identification of the new, multiple RV-C necessitated a better way to keep track of isolates. The sequences clearly indicated that historical recombination was rampant as these species developed, as it was for other Enteroviruses (Huang *et al.*, 2009; Ledinko, 1963; McIntyre *et al.*, 2010). At least one-third of extant rhino isolates apparently arose this way ('stars' on Fig. 4.1), swapping large portions of their polyproteins and 5′ untranslated regions (UTRs) (Palmenberg *et al.*, 2010). Surprisingly, though, there are very few examples of intra-capsid recombination. Presumably these proteins need to act as a functional unit where segment swapping could be detrimental.

Citing the successful use of a sequence-based *Enterovirus* classification system, the ICTV Picornavirus Study Group recommended that rhinovirus genotypes also should be defined by sequence identities in their capsid proteins. Taking the lead in these studies, Chloe McIntyre, Nick Knowles and Peter Simmons did a heroic job collating and sorting reams of existing sequences, including all (then) available RV-C (McIntyre *et al.*, 2013). They proposed that nucleotide identities (< 13% diversity) in the VP1 capsid gene should brand similar isolates in an RV genotype. In the absence of such information, preliminary pseudotyping also can be estimated (guesstimated) from VP4–VP2 data (< 10% diversity). Once a genotype is established with adequate sequence depth from clinical isolates, typing of similar strains is more quickly accomplished by 5′ internal ribosome entry site (IRES)-region sequencing of PCR amplicons, supplemented by capsid data, when needed (Bochkov *et al.*, 2014). Flexibility in the choice of regions allows clinical isolates to be quickly identified with reasonable accuracy and convenient PCR primers. When new strains pop up in such analyses, as currently one or two do each year, the assignment of new genotypes requires at least a full capsid sequence (and preferably a full genome) that is phylogenetically distinct. Proposals for new *Enterovirus* species need to consult the Study Group for a proper genetic perspective.

As currently constituted the RV-A, RV-B and RV-C include (to date) ≈166 genotypes (Fig. 4.1). The preferred nomenclature (McIntyre *et al.*, 2013) designates the species letter

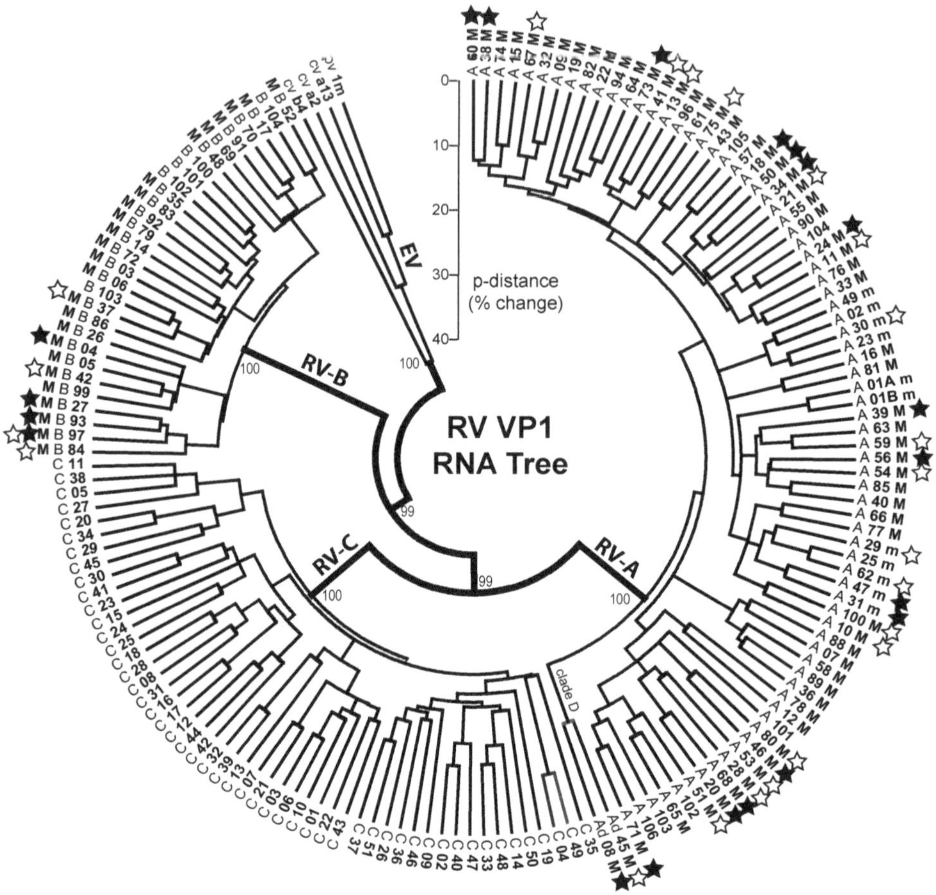

Figure 4.1 Neighbour-joining tree of known RV genotypes according aligned VP1 sequences was calculated with neighbour-joining methods (MEGA) from aligned, VP1 RNA sequences, and rooted with data from four Enteroviruses (EV) of the EV-A, EV-B and EV-C species, similar to Palmenberg *et al.* (2009). The Major ('M', ICAM-1) and minor ('m', LDLR) receptor groups are indicated for the classical 99 RV, if determined experimentally. The newly identified RV-C receptor, CDHR3, is presumed (but still unproven) to be ubiquitous to all isolates in this species. Bootstrap values (per cent of 200 replicates) are indicated at key nodes. Full genome sequences from the original study (Palmenberg *et al.*, 2009) defined known recombinants (★) and parental contributors to those recombinants (☆).

(A, B or C), and type number (e.g. A16). Strain designations are unique to each GenBank accession number and rarely indicated unless required for clarity. The revised taxonomy necessitated a few rearrangements for some of the classic types as duplicates and inconsistencies were pruned out or consolidated. For example, types A44, A95 and A98 were eliminated and these isolates combined into types A29, A8 and A54, respectively. Other types were split (e.g. B52 into B52 plus B104), or their isolates rearranged (e.g. A36 and A89). Type A87 was reassigned to the enterovirus (EV) D species, as EV-D68 (Savolainen *et al.*, 2002). None of the 'retired' numbers are being reused. For the rest of the classic 99, their original type numbers remain unchanged and unique. New additions to each species will from now on be assigned consecutively (e.g. A102 and B102). Presently, the totals for each species depend upon how some popular recognized subtypes are tabulated (e.g. A1a and A1b), but generally there are now about 80 recognized RV-A, 32 RV-B and 55 RV-C.

The mean polyprotein sequence differences (amino acids, aa) among known isolates within any given species are 12–22%, as calculated relative to that species' consensus. The RV-A and RV-B are on the lower side of those values, with the RV-C showing a bit more diversity. Within Enteroviruses as a genus, cut-off values of > 25–30% aa identity in the P1, 2C and 3CD regions usually delineate isolates in different species. To arrive at such values, one has to look at and tabulate a lot of sequence data. The Study Group (Nick Knowles) maintains a superb website listing published sequences (www.picornaviridae.com/) as well as the current ICTV dictates relevant to all picornavirus taxonomy. Note though, that clinical studies, useful as they may be, which generate a myriad of deep but typically short sequence fragments, may not necessarily be listed on that site unless an isolate is a prototype for a new genotype. Keeping track of such commonalities and extracting real comparative information from any 'omics' data requires excellent alignment techniques or access to such. My personal dataset for the composite RV, just counting sequences with 'big pieces' (> 2000 bases each), runs to ≈4 million nucleotides. In the remainder of this article, I will try to show some examples of how access to such information, properly mined, can be very useful in understanding the molecular nuances of rhinoviruses, particularly for the elusive RV-C.

Making an informative alignment

It should be axiomatic that if you want to extract biologically relevant information from a sequence alignment, you must start with a biologically relevant alignment. Yet students in my bioinformatics classes seem mind-numbingly content to hit 'default' at every opportunity with every confronted algorithm and then swallow without question the output. Computers are not the Oracles of Delphi spewing the truth of god. Despite the best intentions of programmers and mathematical biologists, the machines will not think for you. Their sole value as research tools is that they will do intensive, complicated math or matrix searches, usually the basis for most useful algorithms, faster than you can do it in your head, and they will spew out the requested answer into a file you can save. Ask a stupid question, you will get a stupid answer. This is especially important when evaluating alignments. No algorithm yet written will voluntarily tell you that your data are nonsense. You can BLAST, CLUSTAL or MUSCLE any series of random, fictional, or scrambled sequences and the programs will give you absolute outputs. They may even pull out entries from the database and tell you these 'look like' your sequence, for what that's worth. Ask to align gene sequences that are not homologues, or not even real, and you will still get an answer. Well, your sequence has an Ala, and the found sequence has an Ala, so that's a match, right? Silly as it seems, the core illusion that the computer knows better than you, or it will correct your intellectual mistakes is at the heart of most poorly performed analyses (and the source of much garbage in the literature). Fortunately, because they are fast, and they remember outputs, comparison scenarios can be run over and over with different (non-default) parameters. To do so helps one recognize signal from noise and ultimately, develop a better picture of true sequence relationships.

Good sequence sets are long, clean, and deep. Full genomes have more information than single genes and really do not require a lot more work to generate. The impetus for determining the 99 classic RV full-length sequences (Palmenberg *et al.*, 2009) was 2-fold; to put the RV-C into end-to-end genome perspective and to document the true species variability in non-structural genes, particularly for the 2A and 3C proteases. Had the study focused only on the capsids, important information would have been missed about IRESes, 3′ ends, 3D and 3A proteins as well. As these sequences were compiled, however, it became clear there was extensive quality variability in many GenBank deposits. For the RV, it was preferable to include all possible iterations and strain variations to increase confidence that sequence changes (or sequencing mistakes) were properly identified. However, in many instances, it was found that the clinical or survey studies generating shorter fragments, did not always trim all cloning artefacts from the 5′ and 3′ ends or from the middles (Liggett *et al.*, 2014a,b,c). These datasets have immense comparative value but when a particular region looks goofy relative to other sequences, it is typically poorly trimmed (and usually out of frame). *Caveat emptor*! The 'junk' regions of poor coverage, wherever suspected, can be deleted or included with, for example, lowercase letters so as not to have improper weight in alignment formation. I prefer my entries be configured with brief, informative names including the GenBank number so they can be traced back to the source, strain and reference if needed (e.g. C15-GU219984).

Alignment packages using CLUSTAL, MUSCLE or related profile algorithms are readily available. MEGA (Molecular Evolutionary Genetics Analysis) is a free suite (Tamura *et al.*, 2011). The Lasergene (DNAStar) SeqEd module is a nice commercial product. Both have colourful intuitive interfaces that help with editing, and both will work on McIntosh or PC units. Each can import sequences directly from GenBank or use personal files, align the data (see below), allow editing of sequences and the alignment, and give outputs in a variety of formats. The bulk of the RV collection fits nicely into two MEGA files (for example http://virology.wisc.edu/acp/aligns/). One has genome-length (big pieces) data and the other has deep VP1-specific data from clinical studies. Data stored as nucleotides can toggle back and forth to encoded proteins if the open reading frames (ORF) are kept intact. It is important to do so. Pull any 10 random RV sequences from GenBank, ask CLUSTAL to align them (using default settings), and guaranteed, there will not be an intact ORF in the output. When starting from raw sequences, multiple alignment algorithms: (1) do a rough estimate of pairwise distances, (2) align the 'closest' sequences first, inserting insertion or deletion (indels, gaps) where needed to maximize matches, then (3), successively add the rest of the sequences, while respecting and/or augmenting the required indels. Step one assumes all compared genes are homologues, there is no recombination, and there are no sequencing mistakes. Step two uses dynamic programming (or word searches) to create and search pairwise matrices for the 'best' fit between the closest sequences according to amino acid or nucleotide comparison tables. If there are multiple equivalent paths for the placement of indels, the program will choose a default, locking that gap (or its mistaken placement), into subsequent rounds. An erroneous placement of foundation gaps only becomes amplified as more sequences are added in step three. There is no (innate) automated attempt to maintain the reading frame or fine-tune the gaps other than user-implemented end-point editing.

For the RV species, which average ≈32% nucleotide divergence over all sequence pairs (maximum, 43% for B3 vs. C9), program defaults might give some semblance of a functional alignment if you chose that route, but it will not be a good one. Deep data mining to extrapolate high resolution analogy needs something more refined. It took about 3 months to build the files from which the RV-C biology could be properly attacked (Palmenberg *et al.*, 2009). Founder data used superimposed hydrogen bonding maps from RV-A and RV-B crystal structures (capsid, 2A, 3C, 3D). This results in amino acids with the same job in a protein being forced into the same alignment column, regardless of their identity. Closely related sequences were added next, respecting the gaps required by the founder sequences. The natural variety in these deeper data then formed a 'profile' that fit successively ever more diverse sequences, with each round training and guiding the addition of the next, in a process similar to hidden Markov modelling (Eddy, 1998). Every indel in the output iterations was eased within the confines of maximum and minimum matrix pathways (high road and low road) to make sure viral cleavage sites, catalytic sites, structural landmarks and sequence similarity were preserved. Reverse translation of the gapped polyproteins relative to the original RNA sequences formed a core ORF dataset with equivalent indels. To add the 5′ and 3′ ends, representative sequences for the dominant RV clades were probed for optimal and suboptimal RNA folding preferences. Analogous structures (IRES, 5′ cloverleaf, 3′ stem) were superimposed, as had been done for the protein hydrogen bonding maps. The foundation was progressively augmented with the remaining sequences. Linkage of the respective fragments gave genome-length alignments, which for every column, maximized analogy (same function) rather than homology (presumed ancestry). The process done right, was not trivial, but produced a product that was highly informative.

Prediction of an RV-C capsid

The RV-C do not grow in standard tissue culture. In the early years after the initial discovery of these isolates, the only available systems for growing RV-C included laborious air–liquid interface culture (ALI) or recombinant RNA transfection methods which at best, produced 1–2 μg of virus, not the milligrams (or preferably tens of milligrams) required for a legitimate X-ray crystallography structure determination (Ashraf *et al.*, 2013; Bochkov *et al.*, 2011; Hao *et al.*, 2012). It was not until 2016, using a specially adapted recombinant sequence (C15a) and transduced CDHR3-expressing cell lines, that sufficient high-quality preparations (≈100 μg) were obtained and later proven feasible for effective cryo-electron microscopy (EM) analyses (Lui *et al.*, 2016). The hiatus, of years spent waiting for these for these materials, was a problem for the field, because the RV-C clearly behave differently than the RV-A and RV-B in disease phenotypes, drug resistance, receptor choice and immunology (Basta *et al.*, 2014a; Bochkov and Gern, 2012). All these parameters are known to be encoded in the capsid structure. Mechanistic explanations for these phenomena were urgently needed, and the clear, solution to the problem required construction a temporary 'omics-based' model that could help with predictive experiments until the real structure could be finally resolved. In retrospect, those exercises now provide an excellent lesson in what protein modelling can, and cannot do because those RV-C computed outputs, when compared to the real thing, held several surprises. Structural biologists, rightfully, do not put much confidence into protein modelling, especially for divergent sequences. The nub of the problem is that it is still very difficult to predict what solvent-exposed loops look like, and for viruses, the loops are where the good stuff is: on the surface of the particle. Some of the biggest future advances in 'omics' are sure to be in this area. Homology (e.g. MODELLER, ROBETTA) (Kim *et al.*, 2004; Shen and Sali, 2006), threading (e.g. I-TASSER) (Roy *et al.*,

2010) and *ab initio* methods are determined, existing efforts to provide users reasonable interim solutions, and indeed these algorithms can give outstanding results if there is a strong degree of similarity between the template sequence of a known structure, and target sequence one wishes to project (i.e. > 90% aa identity). The more closely related a target is to putative templates in the PDB database, the more unit fragments can be assembled, logically and realistically.

At the time the RV-C model was computed, the PDB recorded ≈35 atomic resolution structures for A1, A2, A16, B3 and B14 protomers. This included numerous derivatives with capsid mutations and drug-bound complexes. But these structures are sufficiently different, that if one attempts a practice modelling as 'proof of concept' the B14 structure (PDB: 4rhv) pretty much fails to predict A16 (PDB: 1aym) on the basis of sequence alone (root mean square deviation, RMSD > 8 Å). Most of the variance (48% capsid aa identity), as anticipated, is in the poorly matched surface loop segments (17% aa identity). Given the even greater diversity of the RV-C, at first, a capsid model seemed unrealistic. But as the refined, deep RV alignments emerged, benchmark residues were identified that unambiguously delineated every internal α or β element of the capsid proteins (Basta *et al.*, 2014b). Between these points for VP1, every RV-C sequence had large striking deletions relative to all other RV-A or RV-B (Fig. 4.2). With analogy at their foundation, the alignments clearly placed these deletions in the βB-βC, βD-βE and βH-βI loops, effectively eliminating whole parts of this important surface protein. When combined, the missing segments make the RV-C VP1 shorter by ≈22 aa relative to the RV-A, and ≈28 aa relative to the RV-B. The change is almost 15% in the protein mass. Outside of VP1 and a few odd discontinuities that are possible sequencing errors, there is no other serious length variance among any other RV-C capsid proteins, VP2, VP3 or VP4. We assumed at the time, that if whole surface segments were missing, one did not need worry about their loop orientations.

VP1 βB-βC loop

```
RV-A con  ./IHISKLDKDHDNDDH--------YNDEGKNFTTW/.
A16       ./IHESVLDIVDN------------YND--QSFTKW/.
RV-B con  ./VHVTEIENKNPTGIEEEGNNMKNNHKEQKLFNDW/.
B14       ./VHVTEIQNKDATGID--------NHREAKLFNDW/.
RV-C-con  ./WANLTLN---------------------NGFKKW/.
C15       ./WGKVTLT---------------------RQYAKW/.
```

VP1 βD-βE loop **VP1 βH-βI loop**

```
RV-A con  ./SEITLV--PCIAAKGDDIGHVVM/../RIVTEEQKHKVEITTRIYHKA/.
A16       ./SEITMV--PSVAAKDGHIGHIVM/../RIVTSEQLHKVKVVTRIYHKA/.
RV-B con  ./SEYTILATASQPNDAQYSSNLSV/../RVVNEHDVHTTLVKIRVYHRA/.
B14       ./SEYTILATASQPDSANYSSNLVV/../RIVNEHDEHKTLVKIRVYHRA/.
RV-C-con  ./MEVTIV----------TNNTGLM/../RALDDTTKND----IKVFVKP/.
C15       ./MEVTIV----------TNNTGLM/../RTLDNSGGND----VKIYVKP/.
```

Figure 4.2 Key indels in RV-C. Protein alignments for composite RV sequences were queried by species for the most frequently occurring amino acid ('con'-sensus) in the vicinity of 3 key VP1 loops that define many surface features 'north' of the capsid canyon region. Gaps ('-') define missing sequence (indels) at that alignment location. For comparison, the reported sequences of A16 (GenBank: L24917), B14 (L05355) and C15 (GU219984) are included.

It took about 3 months to develop and evaluate a series of optimal models for C15 (Basta *et al.*, 2014b), a preferred sequence since there was a cDNA with which to test predictions, and eventually, to determine the actual structure. As with alignment methods, we found the program defaults were rarely successful. In total, 56 C15 models based on a multitude of templates (B14, A16, with or without bound-drugs) contributed to the dataset. At least 3 different prediction methods (I-TASSER, MODELLER, ROBETTA) were used to confirm essential findings. 'Omics' predicted the capsids of RV-C were significantly different from the other known RV structures because of conformations forced by those essential, species-conserved indels. The predominant loss of mass is around the 5-fold axes, significantly changing the surface topology for all features 'north' of the 'canyon' (Fig. 4.3). This includes the canonical RV-A and RV-B receptor binding platforms for ICAM-1 and LDLR, explaining why the RV-C do not interact with these proteins. The interior VP1 drug binding pocket commonly targeted by antivirals is intact in the RV-C, but there are crucial amino acid changes unique to these viruses that could easily predict drug failures in clinical trials (Basta *et al.*, 2014a). Moreover, because the alignments were based on residue-by-residue analogy, it was possible to project a blockage caused by specific, conserved bulky aa substitutions, that could putatively prevent any drug entry, let alone binding as required for efficacy. In toggling between the model structure and its sequences, the total 'omics' dataset clearly suggested possible several possible mechanistic reasons for why classic anti-capsid drugs do not work for the RV-C.

Looking back now, it is actually a bit satisfying that we got some elements of these essential points correct. The eventual cryo-EM structure of C15a (Lui *et al.*, 2016) was templated on this model's PDB, an effort which saved considerable time in backbone and side-chain

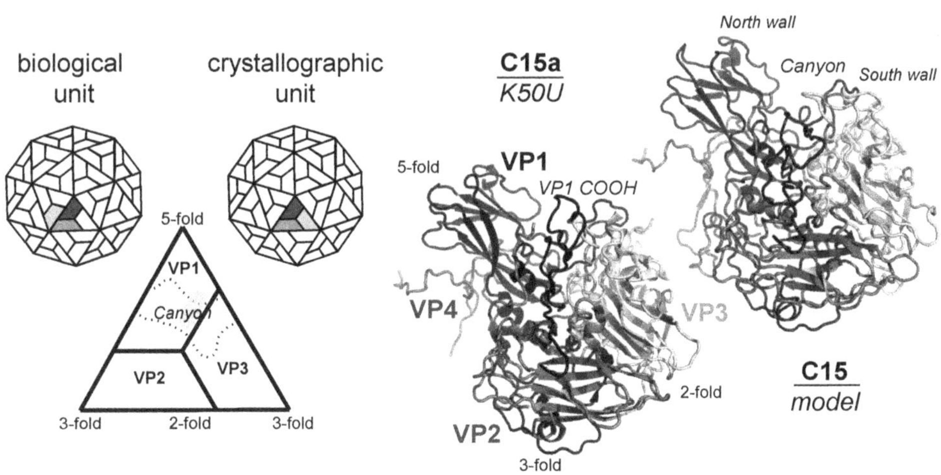

Figure 4.3 Capsid arrangement of rhinoviruses. Icosahedral capsid shows the symmetry of proteins resolved by crystallography, relative to the proteins contributed by a single biological P1 protomer. Triangular close up highlights the canyon feature that encircles each 5-fold axis, particularly for the RV-A and RV-B. The protein structure of a C15a protomer (PDB: 5K0U) highlights the interlocking nature of capsid proteins and the complex of surface loops that contribute to receptor binding and immunogenicity. Next to this, the 'omics' modelled structure of C15, clearly miss-predicted the VP1 COOH segment orientation.

assignments (see Fig. 4.3 for comparison). Predictions about significant missing mass at the 5-fold vertices and the blocked drug-binding pocket held true for the real structure too. In fact, the comparative RMSD between the 'best' model and the real structure (at 2.8 Å, computed over 697 backbone α-carbons was only 1.269 Å, a pretty decent overall fit, considering some terminal bits of the real proteins exhibit disorder and could not be resolved by cryo-EM. The VP2 (RMSD of 1.277 Å) and VP3 (RMSD of 0.858 Å) predictions were particularly good, but then, those RV-C sequences have only very minor aligned indels, relative to the model's template(s). The structure surprises and the places we got things wrong, were (as expected) created by the length variations inherent to the VP1 proteins. Indeed, the VP1 indels in the β-strands change the shape of that protein, particularly in the 5-fold region. What we completely missed though, was that all RV-C, through extensive sequence variation near the COOH terminus of this protein, rearrange this segment (8–10 aa) into an upright finger-like projection, creating 60 new, absolutely unique loops on the particle surface. The RV-A and RV-B tuck this segment flat against the virion surface, and our modelling algorithms, following that plan, did likewise for the RV-C. In reconsideration, I probably should have paid more attention to what the alignments were recording, namely that segment of VP1, displays variation characteristic of the RV-A and RV-B surface immunogenic epitopes (Table 4.1) and logically then, it should not have a buried configuration. Although we got ≈95% of the C15 capsid reasonably modelled, that other incorrect 5% contributes major, novel properties to the particle. The lesson is, as any structural biologist would tell

Table 4.1 Surface sequence variation versus isolate type [reproduced from reference (Basta *et al.*, 2014b)]

		Observed sequence 'Words'[1]			'Words' predict RV type[2]		
Segment	B14 residues	RV-A[3]	RV-B[4]	RV-C[5]	RV-A[3]	RV-B[4]	RV-C[5]
VP2 Nim2(a)	2135–2143	115	21	32	85%	80%	94%
VP2 Nim2(b)	2155–2167	116	38	38	96%	93%	97%
VP3 Nim3(a)	3055–3064	106	36	36	92%	96%	97%
VP3 Nim3(b)	3071–3079	103	31	35	98%	95%	95%
VP3 COOH	3227–3236	68	15	36	95%	82%	94%
VP1 Nim1ab	1083–1095	115	43	0	90%	83%	(deletion)
VP1 Nim1a	1135–1144	76	33	6[6]	85%	95%	61%[6]
VP1 βG	1186–1198	12	10	7	66%	68%	62%
VP1 βG-βH	1205–1212	77	22	36	90%	77%	98%
VP1 COOH	1279–1289	113	39	38	86%	93%	95%
3D-pol	318–331	10	12	21	64%	72%	76%

[1]Unique sequences (words) in same alignment segment.
[2]Words conserved within, but not between types (per cent).
[3]208 isolates, 79 types.
[4]74 isolates, 30 types.
[5]67 isolates, 32 types.
[6]7 of 10 residues deleted.

you, 'omics' can only carry you so far! After that, you really have to do the actual experiment, or in this case, the structure.

Statistical prediction of immunogenicity

I am asked frequently by the lay public why there is not an effective vaccine for the 'common cold'. Ignoring for the moment that ≈20% of 'colds' are from adenoviruses, coronaviruses or RSV, it still can be hard to explain to a non-'omics' person what Nature has done with native RV sequences to make them such perfectly adapted pathogens. Unlike polio or influenza, for which there are successful vaccines, the RV do not emerge in immunologically distinct waves to sweep the world at timed intervals. Properly used, vaccines can optimistically stay ahead of that sort of aetiology. Instead, most of the identified 166 RV types probably circulate continuously with only a minimal 'herd immunity' impetus for evolutionary displacement. In a series of recent studies from the University of Wisconsin, 77 distinct RV genotypes were isolated from subjects at one hospital alone, over the course of only 6–10 years. These included 43 RV-A, 13 RV-B, and 21 RV-C (Liggett *et al.*, 2014a,b,c). The abundance of these types from a single location means it probably is not exposure, but something like discontinuous receptor display that regulates and contributes to the frequency of infection. ICAMs, the RV-A and RV-B dominant receptors, are 'help me' signals up-regulated when normal cells receive a physical or chemical (e.g. cytokine) insult (Chang *et al.*, 2008). When your mother said, 'Don't do that! It will give you a cold,' what she really meant was, 'Don't do that! You will up regulate ICAM-1 and allow the currently resident RV-A and RV-B to infect.' No doubt, as the new CDHR3 RV-C receptor is more closely examined, differential display patterns dependent on host age, physical insult to pulmonary epithelia (smoking? cold temperatures? stress? allergies?) and the alleles inherent to specific individuals (Bochkov *et al.*, 2015; Bonnelykke *et al.*, 2014) will also become better understood.

On top of all this, there is likely a dose effect for exposure that can perhaps override any temporary paucity of receptor display. Someone sneezing directly in your face or the inglorious chore of providing primary care for a nose-dripping infected child are sure predictors of dose-dependent vulnerability, probably for mixed populations of high-titre virus. The RV readily recombine, which means they readily co-infect (Palmenberg *et al.*, 2009). If this is true, universal protection against the RV will require eliciting extensive cross-neutralization over many, many genotypes. For polio, which infects through the gut, achieving protection requires a systemic immunity elicited by vaccination of live/dead virus administered in the proper combinations to cover the three types. For the RV, which grow in the lungs and nasal epithelium, similar systemic immunity would presumably be required, as well as a mucosal immunity component much harder to induce and which may require multiple exposures to all of those continuously circulating types. The live versus dead virus concept in vaccine formulation becomes moot when it is considered there are no known naturally attenuated strains of RV that can induce broad immunity, and no suitable animal test model in which they could be developed. In addition to all this, it is important also to parse what 'omics' teaches us about RV capsid variability.

Immunogenicity and the prediction of immunogenic sites are prime applications of refined sequence analysis. For the RV, the residues contributing to antibody neutralizing immunogenic sites (Nims) were mapped in detail onto the B14 capsid surface by monoclonal antibody (mAb) escape mutations shortly after the initial capsid structure resolution

(Sherry *et al.*, 1986). This seminal study was among the first to describe how viruses evade neutralization by small changes in relevant surface-exposed residues. Similar mappings with other picornaviruses, notably polio (Minor *et al.*, 1986; Nobis *et al.*, 1985; Schild *et al.*, 1983), FMDV (Bolwell *et al.*, 1989; Pfaff *et al.*, 1988) and *Mengo* (Boege *et al.*, 1991) were also carried out but not necessarily to saturation mutagenesis (i.e. all possible neutralizing residues). Since vaccines are the 'Holy Grail' in picornavirus prevention, there is certainly high current interest in predicting those epitopes in other viruses (especially the RV-C) that might make effective universal coverage targets. Unfortunately, the literature is replete with failed attempts to make efficacious peptide-based vaccines as predicted by conserved sequences, particularly for the rhinos (DiMarchi *et al.*, 1986; McCray and Werner, 1987; Palmenberg, 1987). The problem, once again, comes down to 'omics'. The applications need to be well grounded in a fundamental understanding of what the collected sequences are telling us.

Assuming a 30 nm particle (Fig. 4.4), the surface of an RV protomer occupies ≈4500 $Å^2$. Long, thin receptors like ICAM-1 reach deep into the narrow canyon circling the 5-fold axes to contact the residues necessary for interaction (Fig. 4.3). The canyon itself covers about 20% of a protomer surface (≈900 $Å^2$), of which only a handful of residues (≈100 $Å^2$) comprise the key interface (Rossmann, 1994). Several of these (B14: P1155, V1217, H1220, S1223) are reasonably conserved (> 95% ea) among ICAM-using RV-A and RV-B (Bochkov *et al.*, 2011). In the successful vaccines for polio, the receptor footprint and neutralizing epitopes overlap, leading to speculation that excessive immunogenic variation for that virus could have led to unstable particles, or particles that do not bind well to their receptors (He *et al.*, 2003). This limitation is presumably why there are only three viable serotypes of polio. The far greater number of RV-A and RV-B serotypes was originally explained by

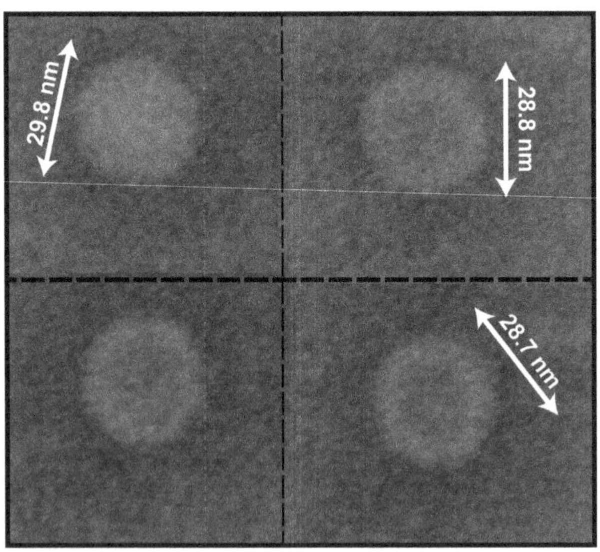

Figure 4.4 EM images of recombinant C15 virus, negative-stained with methylamine tungstate. Arrows show measured diameters of individual particles. Images courtesy of Theodor F. Griggs and Randall Massey, University of Wisconsin, SMPH, EM Facility.

speculating these viruses hide their conserved receptor sites deeper in the canyon where they are sterically inaccessible to antibodies (Rossmann, 1989). The 'canyon hypothesis' further predicted that if these conserved residues in the canyon or nearby could be targeted, a vaccine might be ubiquitously effective.

This approach however, has yet to work. Cryo-EM resolution of B14-bound neutralizing antigen-binding fragment of antibody (contains the paratope) (Fab) fragments (reactive fragments of mAbs) maps a typical paratope footprint at ≈550 Å^2, or 5× larger than the receptor footprint (Smith *et al.*, 1993). Within this, almost any capsid residue substitution has the potential to change the dissociation constant (K_D). The determined structure showed the Fab, but not the virus is conformationally flexible, and actually binds very deeply into the canyon (Smith *et al.*, 1996). But the binding still requires extensive matched surfaces for tight interactions. The problem in achieving universal neutralization (i.e. for more than 1 serotype) is not a question of properly targeting the few conserved surface residues. It's a more fundamental problem created by the surrounding Fab footprint and the native variability over that whole area. The RV have played this game with evolution for thousands of years. The net result is 166 known genotypes, none of which are cross-neutralizing, as proven by the classic 99 serotypes, identified exactly because of these properties.

The sequences record the remarkable extent of capsid surface variability per isolate and per genotype, above and beyond that of standard mutational fixation rates that shaped the remainder of the genomes. From the crystal structure of B14 one can extract the inclusive list of short surface loops, whose sequences contribute to the experimentally mapped epitopes. The alignments then are used to identify the counterparts used in each genotype sequence contributing to these same 'words' (Basta *et al.*, 2014b). If chosen properly, the lexicon of surface words parse almost exactly into the known serotypes (genotypes) of the RV-A and RV-B since these were originally defined by their immunogenicity (Table 4.1). The 'math' for doing this is actually quite simple if one turns the protein alignments sideways in Excel, and then queries along rows or linear segments (analogous residues) for commonalities. As recorded in the sequences, the surface Nim loops (8–12 aa) for the RV-A and RV-B segregate with and predict genotype (serotype) cross-reactivity for all known isolates with a correlation of 92–98%. Non-surface regions of the capsid or segments from non-structural proteins tabulate much lower (≈60–70%), with similar analyses, indicating the legacies of divergent lineages but not the further selective pressures of immunology. All extant RV-C sequences follow similar segregation patterns for the currently defined genotypes. Moreover, although the species-wide VP1 deletions eliminate the mapped analogues to B14 Nim1a and Nim1b, the known sequences suggest that these viruses instead have enhanced variability, perhaps compensatory, in other regions of the surface, notably in the VP1 FMDV loop (βG-βH) and at the COOH-terminus of VP1. Actually, the surface loops of the RV-C record even more diversity than the other species (95–99% correlation with genotype), including for the VP2, VP3 epitopes, Nim2 and Nim3.

What this means, in total, is that 'omics' predicts no antibody [mAb or polyclonal antibody (Ab)] that recognizes any RV, will have a footprint that does not include multiple, highly variable sequences relative to that species. Neutralization for any given genotype (serotype) is possible, but it is unlikely there will be any, much less extensive, cross-neutralization. Nature has already balanced the competing evolutionary needs for capsid stability, common receptor interactions, and immunological diversity. For the RV, surface diversity as the phenotype of relentless immune pressures, was clearly the dominant response. To take out the common cold, we may have to find other targets.

Other uses for RV sequences

This chapter has not dealt with the concept of phylogenies other than to comment on the pairwise VP1 distance values as the basis for the current taxonomy. There are excellent reviews on this subject, particularly from Peter Simmonds' group, as cited above (McIntyre *et al.*, 2013). Within species, or even within most picornavirus genera, the gene layout, including the 5′ and 3′ UTRs, are homologous, meaning they share common ancestry. However, when switching among genera, it must be remembered that for 2A genes, Leader proteins (e.g. in *Cardioviruses* and *Aphthoviruses*), as well as many 5′ and 3′ functional motifs, individual units were probably swapped in by recombination during some point in evolution and, therefore, are not homologues. As with alignment algorithms, phylogeny programs work on the assumption that the sequences you are feeding them are somehow related in evolution, and thus, the 'relationships' relative to putative ancestors can be approximated. Calculations with full-length mixed genus picornavirus genomes do not meet these criteria and will be biased by the regions without common ancestry. It is an almost certainty you will obtain different 'trees' when comparing different genes, and this is why it is essential to report your exact methods and alignments when drawing conclusions from such trees. For the RV, no matter what trees are generated (e.g. Fig. 4.1), and how hard you study them, it becomes quickly obvious that one cannot realistically retrace the ancestral origins of individual genotypes. In a practical sense, there is no obvious gain in doing so. Generally, all known isolates segregate into 3–5 major clades within each species, for every gene, except where recombination has mixed them up. The data for all genes converge on the idea that the RV-A and RV-C are more closely related to each other, than either is to the RV-B, but one would be hard pressed to describe which is more primordial (i.e. the founder), or where the branch point from the other EV species developed (Palmenberg *et al.*, 2010).

As a last point relative to the future of RV 'omics', the curation of clinical sequence has been, and will continue to be essential to disease diagnostics. Some commonalities in the 5′ UTR (IRES) are now recognized as ubiquitous to all RV, and provide excellent starting points for gene chip or PCR genotype recognition techniques (Bochkov *et al.*, 2014). As the biology of the RV-C are sorted out, the more pathogenic strains may reveal themselves through such analyses. Since the RV-C are refractive to capsid-binding drugs, which are otherwise effective on RV-A and RV-B (Basta *et al.*, 2014a; Bochkov *et al.*, 2011), clinical differentiation among the species will be crucial to developing appropriate treatments.

Acknowledgements

The RV efforts in the Palmenberg group are supported by NIH grant U19 AI104317. The author thanks Dr Jim Gern for helpful suggestions and critical reading of the manuscript

Programs and databases (compiled 2015)

BLAST: basic local alignment search tool; blast.ncbi.nlm.nih.gov/Blast.cgi/
CLUSTAL: www.clustal.org/
GenBank: www.ncbi.nlm.nih.gov/genbank/

I-TASSER: iterative threading assembly refinement; https://zhanglab.ccmb.med.umich.edu/I-TASSER/
LaserGene: www.dnastar.com/
MEGA: molecular evolutionary genetics analysis; www.megasoftware.net/
Microsoft Excel® (Microsoft Corporation, Redmond, WA, USA): spreadsheet software; Microsoft Office
MODELLER: salilab.org/modeler/
MUSCLE: multiple seq comparison by log-expectation; www.ebi.ac.uk/Tools/msa/muscle/
PDB: protein data bank; www.rcsb.org/pdb/
Picornavirus Database: www.picornaviridae.com/
Rhinovirus alignments: virology.wisc.edu/acp/aligns
ROBETTA: robetta.bakerlab.org/

References

Adams, M.J., King, A.M.Q., and Carstens, E.B. (2013). Ratification vote on taxonomic proposals to the International Committee on Taxonomy of Viruses. Arch Virol. *158*, 2023-2030. https://doi.org/10.1007/s00705-013-1688-5

Ashraf, S., Brockman-Schneider, R., Bochkov, Y.A., Pasic, T.R., and Gern, J.E. (2013). Biological characteristics and propagation of human rhinovirus-C in differentiated sinus epithelial cells. Virology. *436*, 143-149. https://doi.org/10.1007/s00705-012-1522-5

Basta, H.A., Ashraf, S., Sgro, J.-Y., Bochkov, Y.A., Gern, J.E., and Palmenberg, A.C. (2014a). Modeling of the human rhinovirus C capsid suggests possible causes for antiviral drug resistance. Virology. *448*, 82-90. https://doi.org/10.1016/j.virol.2013.10.004

Basta, H.A., Sgro, J.-Y., and Palmenberg, A.C. (2014b). Modeling of the human rhinovirus C capsid suggests a novel topgraphy with insights on receptor preference and immunogenicity. Virology. *448*, 176-184. https://doi.org/10.1016/j.virol.2013.10.006

Bochkov, Y.A., and Gern, J.E. (2012). Clinical and molecular features of human rhinovirus C. Microbes Infect. *14*, 485-494. https://doi.org/10.1016/j.micinf.2011.12.011

Bochkov, Y.A., Grindle, K., Vang, F., Evans, M.D., and Gern, J.E. (2014). Improved molecular typing assay for rhinovirus species A, B, and C. J Clin Microbiol. *52*, 2461-2471. https://doi.org/10.1128/JCM.00075-14

Bochkov, Y.A., Palmenberg, A.C., Lee, W.-M., Rathe, J.A., Amineva, S.P., Sun, X., Pasic, T.R., Jarjour, N.N., Liggett, S.B., and Gern, J.E. (2011). Molecular modeling, organ culture and reverse genetics for a newly identified human rhinovirus C. Nature Medicine. *17*, 627-632. https://doi.org/10.1038/nm.2358

Bochkov, Y.A., Watters, K., Ashraf, S., Griggs, T.F., Devries, M.K., Jackson, D.J., Palmenberg, A.C., and Gern, J.E. (2015). Cadherin-related family member 3, a childhood asthma susceptibility gene product, mediates rhinovirus C binding and replication. Proc Natl Acad Sci USA. *112*, 5485-5490. https://doi.org/10.1073/pnas.1421178112

Boege, U., Kobasa, D., Onodera, S., Parks, G.D., Palmenberg, A.C., and Scraba, D.G. (1991). Characterization of Mengo virus neutralization epitopes. Virology. *181*, 1-13.

Bolwell, C., Clarke, B.E., Parry, N.R., Ouldridge, E.J., Brown, F., and Rowlands, D.J. (1989). Epitope mapping of foot-and-mouth disease virus with neutralizing monoclonal antibodies. J Gen Virol. *70*, 59-68. https://doi.org/10.1099/0022-1317-70-1-59

Bonnelykke, K., Sleiman, P., Nielsen, K., Kreiner-Moller, E., Mercader, J.M., Belgrave, D., den Dekker, H.T., Husby, A., Sevelsted, A., Faura-Tellez, G., *et al.* (2014). A genome-wide association study identifies CDHR3 as a susceptibility locus for early childhood asthma with severe exacerbations. Nature Genetics. *46*, 51-55. https://doi.org/10.1038/ng.2830

Chan, S.C.H., Shum, D.K.Y., Tipoe, G.L., Mak, J.C.W., Leung, E.T.M., and Ip, M.S.M. (2007). Upregulation of ICAM-1 expression in bronchial epithelial cells by airway secretions in bronchiectasis. Resp Med. *102*, 287-298.

DiMarchi, R., Brooke, G., Gale, C., Cracknell, V., Doel, T., and Mowat, N. (1986). Protection of cattle against foot-and-mouth disease by a synthetic peptide. Science. *232*, 639-641.

Dominguez, S.R., Briese, T., Palacios, G., Hui, J., Villari, J., Kapoor, V., Tokarz, R., Glodé, M.P., Anderson, M.S., Robinson, C.C., *et al.* (2008). Multiplex MassTag PCR for respiratory pathogens in pediatric nasopharyngeal washes by conventional disgnostic testing shows a high prevalence of viruses belonging to a newly recognized rhinovirus clade. J Clin Virol. *43*, 219-222. https://doi.org/10.1016/j.jcv.2008.06.007

Eddy, S.R. (1998). Profile hidden Markov models. Bioinformatics. *14*, 755-763.

Hamparian, V.V., Colonno, R.J., Cooney, M.K., Dick, E.C., Gwaltney, J.M., Hughes, J.H., Jordan, W.S., Kapikian, A.Z., Mogabgab, W.J., Monto, A., *et al.* (1987). A collaborative report: Rhinovirus - Extension of the numbering system from 89 to 100. Virology. *159*, 191-192.

Hao, W., Bernard, K., Patel, N., Ulbrandt, N., Feng, H., Svabek, C., Wilson, S., Stracener, C., Wang, K., Suzich, J., *et al.* (2012). Infection and propagation of human rhinovirus C in human airway epilethial cells. J Virol. *86*, 24-32. https://doi.org/10.1128/JVI.02094-12

He, Y., Mueller, S., Chipman, P.R., Bator, C.M., Peng, X., Bowman, V.D., Mukhopadhyay, S., Wimmer, E., Kuhn, R.J., and Rossmann, M.G. (2003). Complexes of poliovirus serotypes with their common cellular receptor, CD155. J Virol. *77*, 4827-4835.

Huang, T., Wang, W., Bessaud, M., Ren, P., Sheng, J., Yan, H., Zhang, J., Lin, X., Wang, Y., Delpeyroux, F., *et al.* (2009). Evidence of recombination and genetic diversity in human rhinoviruses in children with acute respiratory infection. PLOS ONE. *4*, e6355. https://doi.org/10.1371/journal.pone.0006355

Kapikian, A.Z. (1967). Rhinoviruses: A numbering system. Nature. *213*, 761-762.

Kim, D., Chivian, D., and Baker, D. (2004). Protein structure prediction and analysis using the Robetta server. Nucleic Acid Res. *32*, 526-531. https://doi.org/10.1093/nar/gkh468

King, A.M.Q., Adams, M.J., Carstens, E.B., and Lefkowitz, E.J., eds. (2012). Virus taxonomy: Ninth report of the international committee on taxonomy of viruses (San Diego, CA: Academic Press, Elsevier).

Lau, S.K., Yip, C.C., Lin, A.W., Lee, R.A., So, L.Y., Lau, Y.L., Chan, K.H., Woo, P.C., and Yuen, K.Y. (2009). Clinical and molecular epidemiology of human rhinovirus C in children and adults in Hong Kong reveals a possible distinct human rhinovirus C subgroup. J Infect Dis. *200*, 1096-1103. https://doi.org/10.1086/605697

Lau, S.K.P., Yip, C.C.Y., Tsoi, H.-W., Lee, R.A., So, L.-Y., Ly, Y.-L., Chan, K.-H., Woo, P.C.Y., and Yuen, K.-Y. (2007). Clinical features and complete genome characterization of distinct human rhinovirus genetic cluster, probably representing a previously undetected HRV species, HRV-C, associated with acute respiratory illness in children. J Clin Microbiol. *45*, 3655-3664. https://doi.org/10.1128/JCM.01254-07

Ledinko, N. (1963). Genetic recombination with poliovirus type 1. Studies of crosses between a normal horse serum-resistant mutant and several guanidine-resistant mutants of the same strain. Virology. *20*, 107-119.

Lee, W.-M., Kiesner, C., Pappas, T., Lee, I., Grindle, K., Jartti, T., Jakiela, B., Lemanske, R.F., Shult, P.A., and Gern, J.E. (2007). A diverse group of previously unrecognized human rhinoviruses are common causes of respiratory illnesses in infants. PLos ONE. *2*, e966. https://doi.org/10.1371/journal.pone.0000928

Liggett, S.B., Bochkov, Y.A., Pappas, T., Lemanske, R.F.J., Gern, J.E., Sengalalay, N., Zhao, X., Su, Q., Fraser, C.M., and Palmenberg, A.C. (2014a). Genome sequences of rhinovirus C isolates from Wisconsin pediatric respiratory studies. Genome Announc. *2*, e00203.00214. https://doi.org/10.1128/genomeA.00203-14

Liggett, S.B., Bochkov, Y.A., Pappas, T., Lemanske, R.F.J., Gern, J.E., Sengamalay, N., Zhao, X., Su, Q., Fraser, C.M., and Palmenberg, A.C. (2014b). Genome sequences of rhinovirus A Isolates from Wisconsin pediatric respiratory studies. Genome Announc. *2*, e00200-00214. https://doi.org/10.1128/genomeA.00200-14

Liggett, S.B., Bochkov, Y.A., Pappas, T., Lemanske, R.F.J., Gern, J.E., Sengamalay, N., Zhao, X., Su, Q., Fraser, C.M., and Palmenberg, A.C. (2014c). Genome sequences of rhinovirus B isolates from Wisconsin pediatric resporatory studies. Genome Announc. *2*, e00202.00214.

Liu, Y., Hill, M.G., Klose, T., Chen, Z., Watters, K.E., Bochkov, Y.A., Jiang, W., Palmenberg, A.C., and Rossmann, M.G. (2016). Atomic structure of a rhinovirus C, a virus species linked to severe childhood asthma. Proc Nat Acad Sci USA. *113*, 8997-9002. https://doi.org/10.1073/pnas.1606595113

Luo, M., Rossmann, M.G., and Palmenberg, A.C. (1988). Prediction of three-dimensional models for foot-and-mouth disease virus and hepatitis A virus. Virology. *166*, 503-514.

Luo, M., Vriend, G., Kamer, G., Minor, I., Arnold, E., Rossmann, M.G., Boege, U., Scraba, D.G., Duke, G.M., and Palmenberg, A.C. (1987). The atomic structure of Mengo virus at 3.0 A resolution. Science. *235*, 182-191.

McCray, J., and Werner, G. (1987). Different rhinovirus serotypes neutralized by antipeptide antibodies. Nature. *329*, 736-738. https://doi.org/10.1038/329736a0

McIntyre, C., L., Knowles, N.J., and Simmonds, P. (2013). Proposals for the classification of human rhinovirus species A, B and C into genotypically assigned types. J Gen Virol. *94*, 1791-1806. https://doi.org/10.1099/vir.0.053686-0

McIntyre, C.L., McWilliam Leitch, E.C., Savolainen-Kopra, C., Hovi, T., and Simmonds, P. (2010). Analysis of genetic diversity and sites of recombination in human rhinovirus species C. J Virol. *84*, 10297-10310. https://doi.org/10.1128/JVI.00962-10

Minor, P.D., Ferguson, M., Evans, D.M.A., Almond, J.W., and Icenogle, J.P. (1986). Antigenic structure of polioviruses of serotypes 1, 2 and 3. J Gen Virol. *67*, 1283-1291. https://doi.org/10.1128/JVI.00962-10

Nobis, P., Zibirre, R., Meyer, G., Kühne, J., Warnecke, G., and Koch, G. (1985). Production of a monoclonal antibody against an epitope on HeLa cells that is the functional poliovirus binding site. J Gen Virol. *66*, 561-564. https://doi.org/10.1099/0022-1317-67-7-1283

Palmenberg, A.C. (1987). Antipeptide antibodies. A vaccine for the common cold? Nature. *329*, 668-669. https://doi.org/10.1038/329668a0

Palmenberg, A.C., Rathe, J.A., and Liggett, S.B. (2010). Analysis of the complete genome sequences of human rhinovirus. J Allergy Clin Immunol. *125*, 1190-1199. https://doi.org/10.1016/j.jaci.2010.04.010

Palmenberg, A.C., Spiro, D., Kuzmickas, R., Wang, S., Djikeng, A., Rathe, J.A., Fraser-Liggett, C.M., and Liggett, S.B. (2009). Sequencing and analysis of all known human rhinovirus genomes reveals structure and evolution. Science. *324*, 55-59.

Pfaff, E., Thiel, H.-J., Beck, E., Strohmaier, K., and Schaller, H. (1988). Analysis of neutralizing epitopes on foot-and-mouth disease virus. J Virol. *62*, 2033-2040.

Rossmann, M.G. (1989). The canyon hypothesis. Hiding the hostcell receptor attachment site on a viral surface from immune surveillance. J Biol Chem. *264*, 14587-14590.

Rossmann, M.G. (1994). Viral cell recognition and entry. Protein Sci. *3*, 1712-1725. https://doi.org/10.1002/pro.5560031010

Rossmann, M.G., Arnold, E., Erickson, J.W., Frankenberger, E.A., Griffith, J.P., Hecht, H.-J., Johnson, J.E., Kamer, G., Luo, M., Mosser, A.G., *et al.* (1985). The structure of a human common cold virus (Rhinovirus 14) and its functional relations to other picornaviruses. Nature. *317*, 145-153.

Roy, A., Kucukural, A., and Zhang, Y. (2010). I-TASSER: a unified platform for automated protein structure and function prediction. Nat Protoc. *5*, 725-738. https://doi.org/10.1038/nprot.2010.5

Savolainen, C., Blomqvist, S., Mulders, M.N., and Hovi, T. (2002). Genetic clustering of all 102 human rhinovirus prototype strains: serotype 87 is close to human enterovirus 70. J Gen Virol. *83*, 333-340. https://doi.org/10.1099/0022-1317-83-2-333

Schild, G.C., Minor, P.D., Evans, D.M.A., Ferguson, M., Stanway, G., and Almond, J.W. (1983). Molecular basis for the antigenicity and virulence of poliovirus type 3. In Modern Approaches to Vaccines, R.M. Channock, and R.A. Lerner, eds. (Cold Spring Harbor, NY: Cold Spring Harbor Laboratory), pp. 27-35.

Shen, M.Y., and Sali, A. (2006). Statistical potential for assessment and prediction of protein structures. Protein Sci. *15*, 2507-2524. https://doi.org/10.1110/ps.062416606

Sherry, B., Mosser, A.G., Colonno, R.J., and Rueckert, R.R. (1986). Use of monoclonal antibodies to identify four neutralization immunogens on a common cold picornavirus, human rhinovirus 14. J Virol. *57*, 246-257.

Smith, R.D. (2006). Responding to global infectious disease outbreaks: Lessons from SARS on the role of risk perception, communication and management. Soc Sci Med. *63*, 3113-3123. https://doi.org/10.1016/j.socscimed.2006.08.004

Smith, T.J., Chase, E.S., Schmidt, T.J., Olson, N.H., and Baker, T.S. (1996). Neutralizing antibody to human rhinovirus 14 penetrates the receptor-binding canyon. Nature. *383*, 350-354. https://doi.org/10.1038/383350a0

Smith, T.J., Olson, N.H., Ceng, R.H., Liu, H., Chase, E.S., Lee, W.M., Leippe, D.M., Mosser, A.G., Rueckert, R.R., and Baker, T.S. (1993). Structure of human rhinovirus complexed with Fab fragments from a neutralizing antibody. J Virol. *67*, 1148-1158.

Stanway, G., Hughes, P.J., Mountford, R., Minor, P.D., and Almond, J.W. (1984). The complete nucleotide sequence of a common cold virus: human rhinovirus 14. Nucleic Acids Res. *12*, 7859-7875.

Tamura, K., Peterson, D., Peterson, N., Stecher, G., Nei, M., and Kumar, S. (2011). MEGA5: molecular evolutionary genetics analysis using maximum likelihood, evolutionary distance, and maximum parsimony methods. Mol Biol Evol. *28*, 2731-2739. https://doi.org/10.1093/molbev/msr121

Viral Population Dynamics and Sequence Space

5

Gonzalo Moratorio and Marco Vignuzzi*

Institut Pasteur, Viral Populations and Pathogenesis Unit, CNRS UMR 3569, Paris, France.

*Correspondence: marco.vignuzzi@pasteur.fr

https://doi.org/10.21775/9781910190739.05

Abstract

Without question, the study of Enteroviruses at the molecular scale has uncovered the vast majority of our knowledge on the biology of these viruses, how they invade and replicate in host cells, the molecular determinants of pathogenesis, molecular targets for vaccines and antivirals. Molecular studies have also helped understand how these viruses propagate as populations, and being RNA viruses, they have considerably contributed to the study of RNA virus population dynamics. Because at this level we are under the broader umbrella of population genetics and evolutionary sciences, it would be impossible to present a uniquely *Enterovirus*-specific chapter on the matter. Rather, we have opted to present an overview of the questions that have been addressed in RNA virus population biology, with the state of the art and future directions, giving special attention to instances in which picornaviruses, and more specifically the Enteroviruses, have made significant contributions. Here, we describe how RNA viruses have come to be viewed as highly diverse populations, how Enteroviruses have heavily contributed to understanding two principal mechanisms by which diversity is maintained (mutation and recombination), the current work attempting to fully characterize the genetic sequence space and the phenotypic fitness landscapes occupied by these viral populations, and several kinds of intra-population dynamics that have been observed and require further study in the future. While *Enterovirus* studies punctuate the field throughout, it is important to bear in mind that in several cases, even more detailed studies have been performed in other virus families, and the reader is encouraged to seek those out for a more complete picture.

Quasispecies dynamics of Enteroviruses: model RNA viruses

Fifty years ago our view of RNA viruses as simple, uniform organisms changed drastically with the first experimental evidence that these viruses replicate with a high error frequency. This evidence was brought forward by Eggers and Tamm in 1965, who examined drug-dependent coxsackie A9 virus populations comprised of variants displaying a variety of phenotypes (Eggers and Tamm, 1965): either drug resistance or varying degrees of drug sensitivity. The authors concluded that this virus must be mutating from one state to the other. These observations remained largely overlooked and were not extended until 1978, when Batschelet, Domingo and colleagues in Charles Weissmann's laboratory demonstrated that Qβ phage populations were of genetically heterogeneous composition (Domingo *et al.*, 1978). This finding was received by the scientific community with more enthusiasm and attention: 'The genome of Qβ phage cannot be described as a defined unique structure, but rather as weighted average of a large number of different individual sequences'. The authors calculated that each round of replication produces progeny that differs by one or two mutations from the parental genome (Batschelet *et al.*, 1976). Enthusiasm for this work was fuelled by the work of Manfred Eigen, who in 1971 formulated the first mathematical treatment linking Darwinian natural selection to the self-organization of primitive entities (presumably RNA molecules), assuming erroneous replication of the templated molecules. The theory was proposed to understand the origins of early life (Eigen, 1971). This mathematical framework was later extended and further developed by Eigen and Schuster to become known as the *quasispecies theory* (Eigen and Schuster, 1977). It described the behaviour of self-replicating macromolecules that due to a high error frequency from lack of proofreading mechanisms, replicate into dynamic distributions of genetically similar molecules that surround a master sequence. With respect to the original formulations, the theory assumes that replication reaches infinite population sizes in equilibrium (Eigen, 1992; Eigen *et al.*, 1988; Epstein and Eigen, 1979). By the late 1970s, the conceptual similarities between RNA virus populations and Eigen's self-replicating molecules became evident, initiating a new framework in which to study RNA virus biology.

Given the deterministic nature of the original theory, it was necessary to incorporate stochastic (chance) events to make the theoretical and mathematical framework fit real systems. To better describe the evolutionary dynamics of RNA viruses, the theory was elaborated by several authors to consider limited population sizes in variable fitness landscapes (non-equilibrium conditions) (Hu and Saakian, 2006; Park *et al.*, 2010; Saakian and Hu, 2006; Saakian *et al.*, 2006; Wilke, 2001a; Wilke *et al.*, 2001). The term 'quasispecies' thus entered more common usage among virologists to describe RNA virus populations (Holland, 2006; Domingo *et al.*, 1995; Perales *et al.*, 2010). In virology, the term refers to the mutant spectrum, a 'cloud' of non-identical, yet related variants, genetically linked to each other since one genome can (re)-generate another through forward or back mutation. The ensemble of variants thus define a consensus sequence that may or may not exist as an actual genome within the population (Domingo, 2006; Perales *et al.*, 2010). Consequently, an important prediction of the theory is that genetic variation, competition or cooperation, and selection do not act on single genomes but rather, on a larger mutant swarm, meaning the population behaves (*quasi-*) like a single unit (*species*) (Domingo, 2006; Eigen and Biebricher, 1988).

In the decades that followed, the general framework of quasispecies was extended to all other RNA viruses, with extensive pioneering work performed by Esteban Domingo on the picornavirus, FMDV (Domingo *et al.*, 1992). While the theory has been tremendously helpful in advancing our comprehension of RNA virus genetics, it remains contested just to what extent RNA virus populations satisfy the required theoretical conditions (Holmes, 2010; Holmes and Moya, 2002; Jenkins *et al.*, 2001): for example, population sizes are not always large enough to assume 'infinite sizes' particularly early in infection or after bottlenecking events; mutation rates are very high, but may still not be high enough under normal growth conditions to favour group selection. Indeed, populations geneticists argue that many of the terms and concepts used by the theory already exist in classic population genetics, and the conditions specific to quasispecies-like behaviour may thus only occur during certain points of infection. Semantics aside, the body of work attempting to understand the population dynamics of RNA viruses has transformed the original, simplistic view of an RNA virus as a singular entity to a more realistic view in which 'wildtype' represents a heterogeneous and dynamic population of slightly different variants. In the process, these virus-specific studies allowed experimentalists to directly address long-standing concepts in classical population genetics, such as the Red Queen hypothesis (Clarke *et al.*, 1994), Muller's ratchet (Escarmís *et al.*, 1996), fitness variations or founder effects (Clarke *et al.*, 1993; Escarmís *et al.*, 1999; Martínez *et al.*, 1991; Novella *et al.*, 1995a; Yuste *et al.*, 2005). In the last decade, efforts to expand the basic quasispecies theory by relaxing its original assumptions and incorporating even more virus-realistic features have been made: for example, complementation during co-infection (Elena *et al.*, 2010; Iranzo and Manrubia, 2012), fitness landscapes with multiple peaks (Saakian *et al.*, 2006; Sardanyés *et al.*, 2014), robustness and neutrality (Sardanyés *et al.*, 2008; Wilke, 2001b), and different modes of replication and epistasis (Sardanyés and Elena, 2011; Sardanyés *et al.*, 2009).

Like other RNA viruses, Enteroviruses also generate highly heterogeneous populations and exploit almost all known mechanisms of genetic variation to ensure their survival. In *Enteroviruses,* point mutation at the nucleotide level is a primary source of genetic variation within a genotype, which is mainly generated through an error-prone, non-proofreading RNA-dependent RNA-polymerase (RdRp) used to replicate their genetic material; while recombination, is a strong driver of evolution creating new combinations of genotypes.

Mutation rates and RdRp fidelity

Enteroviruses, as other RNA viruses, share a powerful potential for adaptation and rapid evolution, which is associated with two key aspects: high mutation rates and low replication fidelity. A number of pioneering works addressing this was performed with members of this genus. One of the first measurements of the mutation rate of an RNA virus was carried out by Eggers and Tamm on coxsackievirus A9. They calculated a rate of 1×10^{-4} using, as a surrogate measure, the transition of this *Enterovirus* from dependence to independence of 2-benzimidazole (Eggers and Tamm, 1965). A number of studies performed on other Enteroviruses, including poliovirus and EV71, using different time scales and genetic markers as measurements find mutation rates ranging from 10^{-2} to 10^{-5} substitutions per nucleotide site per replication cycle (Belshaw *et al.*, 2011; Drake and Holland, 1999; Jenkins *et al.*, 2002; de la Torre *et al.*, 1990, 1992; Sanjuán, 2012; Sedivy *et al.*, 1987; Vignuzzi *et al.*, 2006). Although the values range across several orders of magnitude, they remain thousands-fold higher than those observed in DNA-based organisms and discrepancies in these values likely result from experimental differences, underscoring the need to normalize data as proposed by Sanjuán (Sanjuán, 2012; Sanjuán *et al.*, 2010, 2012).

The high error rates and low fidelity of viral RNA-dependent RNA polymerases have been considered a consequence of selection for faster replication – a sloppy enzyme that makes progeny more quickly and more abundantly could outcompete a slower one. Alternatively, theory also predicts that higher mutation rates could be favoured by selection in a dynamic environment requiring continuous adaptation. The answer, is likely a composite of both, where fast replicators with higher error rates may have been optimized by natural selection

(Elena and Sanjuán, 2005; Holmes, 2003). Indeed, studies on poliovirus and coxsackievirus polymerases show that elongation speed and fidelity are tightly linked (Campagnola *et al.*, 2015). For RNA viruses, low replicative fidelity generates a diverse population of variants, and although the majority will generally be less fit, a few are expected to have the potential to emerge and dominate in a sudden change in environment. On the contrary, a homogeneous population resulting from high replicative fidelity could fare well under optimal growth conditions, but would lack this flexibility of adaptation in a dynamic host environment.

Once again, the Enteroviruses have helped address these hypotheses. The first experimental support was provided by two independent series of studies following the isolation of the first high-fidelity variant of an RNA virus polymerase, the poliovirus G64S (Pfeiffer and Kirkegaard, 2003; Vignuzzi *et al.*, 2005). This variant was obtained by serially passaging poliovirus in the presence of the RNA mutagen ribavirin. This base analogue is mistakenly incorporated into the genome by the viral RdRp and mistemplates in subsequent replication cycles, resulting in the accumulation of lethal mutations (Crotty *et al.*, 2000, 2001, 2002). This ribavirin-resistant variant has a single amino acid substitution (G64S) in the viral polymerase that increases fidelity by approximately 5-fold, decreasing the likelihood of misincorporating ribavirin as a base (Arnold *et al.*, 2005). Additionally, assays for selectable markers indicated that this virus has a lower mutation rate and its population exhibits less genetic diversity (Pfeiffer and Kirkegaard, 2003; Vignuzzi *et al.*, 2006). Importantly, although the G64S variant presents wildtype-like replication rates in constant environments, such as in immortalized cell lines, it is attenuated *in vivo* and manifests restricted tissue tropism. These studies demonstrate that maintaining lower fidelity and higher mutation rates is beneficial to an RNA virus, such as poliovirus, that needs to respond to different selective pressures *in vivo* (Pfeiffer and Kirkegaard, 2005; Vignuzzi *et al.*, 2005, 2008). Along the same lines, by benefiting from the structural knowledge of the G64S poliovirus variant, high-fidelity variants of human enterovirus 71 (EV71), G64R and G64T, were identified (Sadeghipour and McMinn, 2013). A G64S variant of the coxsackievirus B3 polymerase was also shown to significantly increase fidelity of the purified enzyme in a biochemical assay, although the virus itself was not viable, possibly due to considerably reduced elongation rates (Campagnola *et al.*, 2014).

Since the initial studies on the *Enterovirus* position 64 variants of the RdRp, using RNA mutagens to identify high- and low-fidelity variants has become a general approach to isolate mutator/antimutator variants of other RNA viruses (Beaucourt *et al.*, 2011). Further studies on poliovirus fidelity uncovered several other RdRp residues involved in both increasing and decreasing fidelity (Campagnola *et al.*, 2014; Korneeva, 2007; Liu *et al.*, 2013), leading to the identification of a 'fidelity network' of residues within the interior of the polymerase involved in nucleotide selection and incorporation (Yang *et al.*, 2010, 2012). An initial study on the mutagenesis of another *Enterovirus*, CVB3, also identified a high-fidelity RdRp variant A372V and a low-fidelity RdRp variant S299T (Levi *et al.*, 2010). An RdRp variant of human EV71, L123F, was also isolated, which increases fidelity and reduces pathogenicity alone or in combination with the G64R mutation *in vivo* (Meng and Kwang, 2014).

Further targeting of residues in CVB3 based on structural similarities and their implication in the closely related poliovirus fidelity network (Arnold *et al.*, 2005; Gong and Peersen, 2010; Thompson *et al.*, 2007) generated nine lower fidelity variants that present elevated mutation frequencies (Gnädig *et al.*, 2012). This panel of mutator variants were also attenuated *in vivo*, and together with the extensive literature on lethal mutagenesis of RNA viruses, demonstrated that there is an upper limit to RNA virus mutation rates (Perales *et al.*, 2011). It should be noted that along with the considerable body of work performed on Enteroviruses described here, these observations have been further validated in other picornaviruses, particularly with FMDV (Arias *et al.*, 2008; Ferrer-Orta *et al.*, 2009; Zeng *et al.*, 2013, 2014), as well as RNA viruses from other families (for review see Beaucourt and Vignuzzi, 2014). In summary, the observations obtained with high- and low-fidelity variants of Enteroviruses reveal that nature has fine-tuned mutation rates to strike a balance between maintaining genetic integrity while generating enough diversity to optimize adaptability, and that the range of fidelity modulation within viable virus populations remains within the same order of magnitude.

Recombination

Although RNA viruses were once thought to experience little recombination, both experimental studies and *in silico* analysis of the viral sequence database show it to occur readily in many virus families and not without consequence. In fact, recombination is linked to increments in host range and virulence, and resistance to immune responses and antivirals (for review see Simon-Loriere and Holmes, 2011). While point mutations drive incremental steps in evolution, recombination also plays a significant role by promoting even larger evolutionary steps through the exchange of larger spans of nucleotide sequences between different genomic RNA molecules. Such events are important mechanisms for producing new genomes with selective growth advantages or compensating and correcting for mutations generated during error-prone replication (Worobey and Holmes, 1999).

Recombination within the *Enterovirus* genus has been phenotypically observed since the 1960s (Ledinko, 1963), and homologous recombination was first described in poliovirus (Copper *et al.*, 1974). Enteroviruses are genetically and antigenically highly variable, human Enteroviruses comprise four species A–D (HEV-A-D), and recombination within and between serotypes considerably contributes to their genetic diversity. Their global prevalence and high infection rates in infants greatly facilitate natural recombination. Moreover, genetic recombination represents a widespread phenomenon at both levels, within (intra-typic) and between (inter-typic) *Enterovirus* types (Lukashev, 2010; Lukashev *et al.*, 2005), and these events often precede the emergence of novel evolutionary lineages of Enteroviruses (McIntyre *et al.*, 2010; McWilliam Leitch *et al.*, 2009, 2010, 2012; Mirand *et al.*, 2007; van der Sanden *et al.*, 2011). However, recombination patterns differ between *Enterovirus* species and between types within species. Recombination within the P1 structural protein region tends to occur between viruses belonging to the same serotype, which by definition have the most similar sequences and capsid structures (Oberste *et al.*, 2004), although this has been documented between serotypes as well (Bouslama *et al.*, 2007; Simmonds and Welch, 2006). The inter-typic recombination break-points are more often located in the 5′UTR and the non-structural protein-coding regions (P2 and P3) (Huang *et al.*, 2009; Schibler *et al.*, 2012; Simmonds and Welch, 2006). The frequency of recombination is higher in HEV-B species than in HEV-A species (Simmonds and Welch, 2006) and very low in HEV-D species (Harvala *et al.*, 2011; Smura *et al.*, 2007). Among this genus, HEV71 is the most studied non-polio *Enterovirus*, of notable clinical relevance in Asia. In a recent study, Lukashev and coworkers showed that the recombination events leading to the emergence of HEV71 subtypes also involved members of the global HEV-A gene pool (Lukashev *et al.*, 2014). Yet upon their emergence, HEV71 subtypes ceased recombining with the rest of the HEV-A gene pool, corroborating perfectly with other phylogenetic analyses recently carried out on different genomic regions (Bible *et al.*, 2008; McWilliam Leitch *et al.*, 2012; van der Sanden *et al.*, 2011). Furthermore, this correlates with the high degree of genetic conservation in HEV71, which recombines five times less frequently than its peers in the A species, CVA2, CVA4 and CVA10 (Lukashev *et al.*, 2014).

A great deal of attention has been brought to the HEV-C group, particularly with the ongoing efforts to eradicate poliovirus. Recombinant strains within this group show they may have incorporated parts of the poliovirus genome (Brown *et al.*, 2003). Moreover, recombined poliovirus strains have been isolated from different sources: healthy vaccines, vaccine-associated poliomyelitis cases and immune-deficient patients who excrete virus for long periods (Combelas *et al.*, 2011; Esteves-Jaramillo *et al.*, 2014; Jegouic *et al.*, 2009; Rakoto-Andrianarivelo *et al.*, 2007, 2008; Riquet *et al.*, 2008; Rousset *et al.*, 2003). These studies illustrate the clinical importance of recombination in viral emergence. From a genetics perspective, recombination contributes to RNA virus population dynamics by empowering viruses to explore a bigger area of sequence space than could be accessible by single point mutation, in a shorter amount of time. Furthermore, recombination allows the exchange of 'functional cassettes' of genetic information that have at least proved viable in their original genetic backgrounds. For example, Xiao and colleagues showed that a recombinant-deficient poliovirus mutant was unable to incorporate beneficial mutations leading to the accumulation of detrimental ones. Subsequently, the authors demonstrated that viral fitness was significantly decreased for this specific mutant and, most importantly, adaptation *in vivo* was restricted compared with the wild type virus (Xiao *et al.*, 2016). Of clear benefit to the virus, this phenomenon complicates the researcher's task of characterizing sequence space, since it shows that the theoretical sequence spaces occupied by two distinct viruses, can abruptly be made to overlap.

Sequence space and fitness landscapes

In evolutionary biology, sequence space is a mathematical representation of all possible combinations for a given genome (nucleotide) or protein (amino acid) sequence. For a given sequence of *n* length that could display any of 4 nucleotides or 20 amino acids at each position, the sequence space can be described as a *n*-dimensional hypercube occupied by up to 4n or 20n distinct sequences or coordinates. Thus, genotypes that are similar in sequence are said to be 'close' to one another, while those that are altogether different are found in distant regions of this space, requiring numerous mutational 'steps' to reach one another. Theoretical sequence space is immensely larger than what is actually occupied by biologically viable genotypes, since most sequence combinations are nonsensical and non-viable at the nucleotide and protein level. Pioneering work by Sanjuán and colleagues using vesicular stomatitis virus revealed that the majority of mutations randomly inserted into the genome are highly detrimental (Domingo-Calap *et al.*, 2009; Sanjuán *et al.*, 2004a,b). Recently, a novel approach based on next-generation sequencing was described to accurately define the mutation rates of poliovirus and uncover insights into the mutational landscape of the viral population (Acevedo *et al.*, 2014). By monitoring changes in variant frequencies on serially passaged populations, fitness values were theoretically determined for thousands of mutations across the viral genome. In addition, these fitness values were mapped onto three-dimensional structures of viral proteins in order to explore structure–function relationships. The distribution of fitness effects for non-synonymous mutations was shown to encompass primarily deleterious mutations, confirming the previous studies on a much larger scale.

The relative viability of individual genotypes could thus be integrated into the concept of sequence space by attributing a fitness value to each genotype, to generate a fitness landscape representing the viable regions of sequence space (Fig. 5.1). In practical terms, sequence space could then be represented (albeit over-simplified) as a two-dimensional map atop which a third axis of fitness values can be used to build a topography where peaks and valleys represent high and low fitness, respectively. The idea of illustrating the distribution of fitness values, i.e. the relative capacity to generate progeny compared to other genotypes, as a landscape was originally introduced by Sewall Wright in 1932. Even today, the concept of fitness landscapes has experienced minimal change. Early on, a principal reason was the simple lack of knowledge on the molecular basis of adaptation, the 'hereditary units' that underlie genotypic space; yet still today, the massive genomes of higher organisms and the difficulty in experimentally measuring individual fitness values renders the task of generating real fitness landscapes nearly impossible. For the time being, the fitness landscape is used to graphically and conceptually represent the relationship between genotypes and phenotypes. In other words, to better understand what genotypes (areas of sequence space) are viable, how evolution progresses as 'walks' along the landscape and how adaptation manifests as 'climbs' to higher positions on the fitness surface (different genotype → different fitness → different evolutionary success). Given their extreme mutation frequencies, short generation times, small genomes and ease of experimental manipulation, RNA viruses such as the Enteroviruses are ideal organisms with which to generate empirical data on sequence space and fitness landscapes.

Regarding the related concepts of sequence space and fitness landscape, one can imagine a scenario where over the countless generations in which an organism faces a finite, albeit complex, set of selective pressures, evolution would favour a genotype that finds itself in a more hospitable region of sequence space where a larger number of neighbouring genotypes bear similar fitness values (e.g. silent mutations or conservative amino acid changes). Using a panel of 48 randomly mutagenized and genetically trackable populations of poliovirus, Lauring and Andino demonstrated that the sequence space surrounding the wildtype poliovirus master sequence is indeed a zone of neutral fitness, where mutations have minimal effect (Lauring and Andino, 2011). Since viruses such as poliovirus require error-prone replication for adaptation (Vignuzzi *et al.*, 2006), yet the majority of mutations are detrimental (Acevedo *et al.*, 2014), it is possible that they have evolved genome sequences that would buffer the effect of mutation to some extent. Indeed, *genetic or mutational robustness* – the constancy of phenotype in light of changes in genotype, has been proposed as such a buffering phenomenon (Elena, 2012; Masel and Siegal, 2009). Being difficult to compare between organisms of different species, whether robustness exists as a biological, and thus inheritable and evolvable, trait is not yet clear. As a first step in this direction, mutational robustness has been confirmed in digital systems (Azevedo *et al.*, 2006; Elena and Sanjuán, 2008). Over the past ten years, a few studies have experimentally addressed this concept in RNA viruses, finding evidence that this buffering allows a viral population to increase its genetic diversity without dramatic alterations in phenotype (Codoñer *et al.*, 2006; Goldhill *et al.*, 2014). One comparison of the sensitivity of poliovirus and coxsackievirus B3 to mutations revealed that

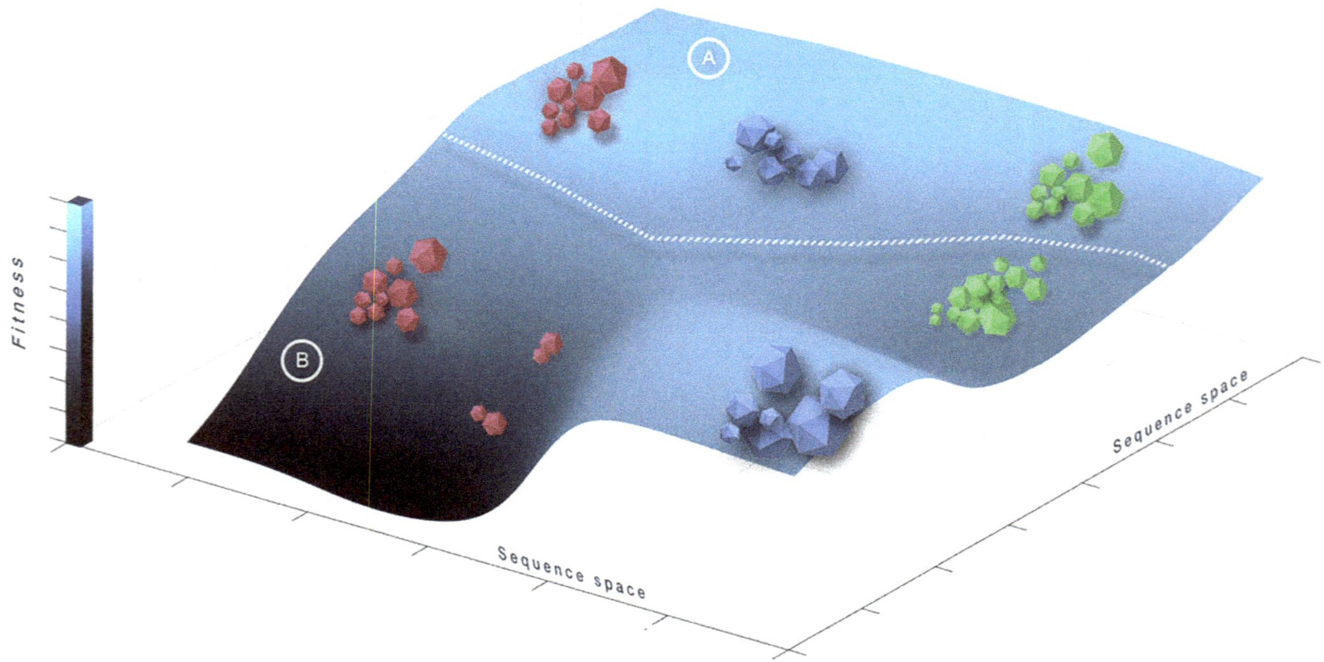

Figure 5.1 Schematic representation of a fitness landscape in action. Three-dimensional landscape, in which sequence space (genotypic space) is represented on the x–y plane and fitness is represented on the z-axis. (A) three different viral populations (red, green and blue) with similar fitness are observed. (B) stress (antiviral or immune pressure) has been introduced in the environment. Different evolutionary trajectories are observed. The green population suffers a fitness loss and moves into a fitness valley remaining viable but at a fitness cost. The blue population faces the introduced changes by exploring other areas of the sequence space with higher fitness – a fitness peak. Finally, the red population drastically falls into a deep valley representing a significant loss in fitness, likely leading to population extinction.

poliovirus is intrinsically more robust than CVB3, and suggests that CVB3 bears a RdRp of higher fidelity to potentially compensate this effect (Graci *et al.*, 2012). In an interesting study comparing poliovirus constructs whose codons had been shuffled without altering amino acid sequence, Lauring and colleagues provided evidence that the mutational robustness of the original wildtype sequence had been reduced and may account at least partially for the observed attenuation *in vivo* (Lauring *et al.*, 2012). However, it is difficult to separate potential impacts on robustness in these constructs from other effects related to codon de-optimization, as other studies on these and other viruses suggest that attenuation is the result of specific codon pair biases introduced during codon shuffling (Le Nouën *et al.*, 2014; Wimmer and Paul, 2011; Yang *et al.*, 2013a). Recently, the relationship between sequence space and mutational robustness was empirically addressed using Enteroviruses as a model. Coxsackievirus B3 genomes were rationally altered in order to redirect their evolutionary trajectories towards detrimental mutations. By rewiring synonymous codons for the amino acids serine and leucine, virus populations were placed in regions of sequence space more likely to lead to stop codons after mutation occurs. Consequently, these variants accrued more stop mutations, leading to attenuation. The proposed genomic design thus conferred a lower mutational robustness by driving these Enteroviruses to 'hostile' regions in sequence space (Moratorio *et al.*, 2017).

In fact, one could envisage that a virus may have evolved to localize in a region of sequence space where mutations would give rise to a maximum of non-conservative mutations. Such mutations, although largely detrimental in the current environment, would result in changes in structure or function that could confer selective advantages when environmental changes considerably alter the fitness landscape and require rapid evolution towards neighbouring fitness peaks. This concept, known as evolvability, is another biological trait that is difficult to directly address experimentally. Although seemingly discordant, recent theory and biology indicate that both robustness and evolvability may be compatible features that are accessed through neutral networks (Goldhill *et al.*, 2014; Hogeweg, 2012). In addition, the relationship between robustness and evolvability might be particularly important for viral pathogenesis (Lauring *et al.*, 2012). It is thus important to understand the link between these two concepts and their definitions, and how the complex cloud of variants generated by an RNA virus fits within these models; but exploring this remains a challenge.

As mentioned above in the specific context of robustness, the most thorough work addressing populations in sequence space and fitness landscapes is theoretical and at best, experimentally tested *in silico* using digital organisms (Wilke *et al.*, 2001a; Wilke, 2001b; Wilke and Ronnewinkel, 2001). Although not biological in nature, the contribution of these studies is significant, showing that in principle the evolution of complex quasispecies populations can be monitored and modelled on fitness landscapes (Wilke *et al.*, 2006; Wilke, 2005). Tight collaboration between computational scientists and biologists will be required to design experiments that can provide the right quantitative data to feed these models – a notable example is the body of work on the population dynamics of VSV being conducted by Novella and Wilke (Novella *et al.*, 2004, 2008; Smith-Tsurkan *et al.*, 2010; Wilke *et al.*, 2004).

Despite these advances, the main obstacle to constructing true empirical fitness landscapes remains the difficulty in generating enough data – how many individual genomes and how many fitness measurements are required to build multiple, or even a single, fitness peak(s)? Given the seemingly intangible number of dimensions required to describe even the smallest of genomes, much less those of higher order organisms, current efforts seek to measure and describe the fitness landscapes of a finite number of genes, assuming that all other genetic sequences do not affect the fitness phenotype being reported. Even so, these efforts are in their infancy, and only a handful of studies have been performed in microbes and other organisms (for review, (de Visser and Krug, 2014)). This approach of examining a predefined and feasible set of alleles was recently applied to HIV. The methodology took into consideration all possible epistatic interactions between seven amino acids in a functional domain of the gp120 glycoprotein, for which fitness calculations were measured (da Silva *et al.*, 2010). In a more extended work, these authors generated fitness maps of 55 tissue-culture passaged populations (20 clones per population) of HIV sequenced across the V1–V2 region of the *env* gene (Lorenzo-Redondo *et al.*, 2014). Sequence space was represented as a two-dimensional matrix based on sequence similarity, with fitness values placed on the third axis.

Viral adaptation dynamics: step-wise walks along the landscape

From the limited empirical data available, it seems that the adaptive walks taken by an organism are more limited than those theoretically available, evoking the ideas of John Maynard Smith that each intermediate adaptive step must retain some functionality and viability (Smith, 1970). An adaptive walk is defined as the step-wise process through which a population gains in fitness in a given environment by fixing beneficial mutations. The dynamics of this process will be affected by a many factors: to name a few, the size and fluctuations of the viral population, the strength of the selective pressure, epistatic effects between mutations, intermolecular interaction (complementation, interference) between individual genomes.

Generally, a gain in fitness occurs when viral populations replicate at high multiplicities of infection and large population size, when the impact of detrimental mutations will be less pronounced (Escarmís *et al.*, 1999; Novella *et al.*, 1995a; 1999). This phenomenon is a consequence of an unrestricted competitive optimization process within the large viral population in a stable environment (Domingo and Holland, 1997). However, when high fitness values are reached, even large population sizes can experience bottlenecks, preventing further fitness increases and resulting in stochastic fluctuations at the fitness plateau (Novella *et al.*, 1999). On the other hand, small population passages, such as by plaque-to-plaque transfer, are more likely to cause genetic bottlenecks that can lead to fitness decreases. During the viral life cycle, bottleneck effects take place frequently. The first study showing a significant decrease in viral fitness undergoing population bottlenecks was carried out with bacteriophage φ6 (Chao, 1990), an observation that was later confirmed with many other RNA viruses (Clarke *et al.*, 1993; Duarte *et al.*, 1992; Escarmís *et al.*, 1996; la Iglesia and Elena, 2007; Novella, 2004; Novella and Ebendick-Corpus, 2004; Novella *et al.*, 1995b; Yuste *et al.*, 1999).

It is thus important to bear in mind that much of the work currently performed to experimentally address RNA virus population structure and dynamics is generated in tissue culture experiments involving large population sizes – with the clear benefits of maximum control of the host environment, number of replication cycles, etc. However, it is evident that *in vivo*, much of a virus' life cycle is occurring under small population size, low MOI and in the presence of bottlenecking events. The transmission of virus between tissues and organs within an infected host, for example, are primary bottlenecking events (Ali *et al.*, 2006; Betancourt *et al.*, 2008; Haaland *et al.*, 2009; Li and Roossinck, 2004; Pfeiffer and Kirkegaard, 2006; Quer *et al.*, 2005; Smith *et al.*, 2008). Infection of mice with poliovirus has been an excellent model to demonstrate the effect of anatomical barriers on RNA virus populations *in vivo*: using a mixture of bar-coded viruses, Pfeiffer and colleagues identified several bottlenecks during poliovirus transmission from inoculated sites to the brain (Kuss *et al.*, 2008; Pfeiffer and Kirkegaard, 2006). They found stochastic effects on which variants disseminate through a bottleneck – how this may impact the likelihood of resulting in a clinical versus subclinical infection remains to be determined.

Intra-population interactions: complementation and interference

How the intra-population interactions that are established among components of a mutant spectrum may enhance or quench the replication of the ensemble is not well understood, but has been observed for many viruses. Seminal experiments showed that the fitness of Qβ phage populations of single clones was lower than the average fitness of the population (Domingo *et al.*, 1978), a phenomenon observed later for VSV (Duarte *et al.*, 1994), suggesting that individual genotypes within the population somehow complement each other. Complementation is defined as an enhancement of viral yield caused by the interaction of virus genes or gene products, for example when a functional protein expressed by one genome enhances the replication or overall fitness of another genome whose corresponding protein is truncated or aberrant. Several examples of complementation have shown that debilitated and defective mutants may be maintained in viral populations (Aaskov *et al.*, 2006; Gelderblom *et al.*, 2008; Moreno *et al.*, 1997). In FMDV, serial passaging of a clone at high MOI resulted in the evolution of a viral population dominated by defective genomes, which were still infectious by complementation (García-Arriaza *et al.*, 2004, 2006). In poliovirus, the higher fidelity G64S variant could not disseminate to the central nervous system in mice. On the other hand, co-infection of the G64S population with wildtype populations allowed successful invasion, suggesting complementation between members of the more diverse wildtype population with members of the more genetically restricted G64S mutant cloud (Vignuzzi *et al.*, 2006). The underlying mechanisms remain unclear, as the authors were unable to identify the specific genotypes involved, using the lower throughput molecular clone sequencing available at the time.

Early experimental evidence for interference was found between components of a viral population in VSV, showing that a mutant spectrum of lower viral fitness was able to suppress the phenotype of a mutant with superior fitness, unless the latter was present above a certain frequency (la Torre and Holland, 1990). Following this work, other *in vivo* and *in vitro* experiments were carried out in different virus–host systems adding additional support to interference and suppressive effects within viral populations. The suppression of neurovirulent poliovirus within vaccine preparations by attenuated poliovirus, was also shown to depend on the relative amount of each (Chumakov *et al.*, 1991). A notable study examining this 'virulence threshold' of *poliovirus* demonstrated that wildtype poliovirus is suppressed by an excess of non-pathogenic variants (Lancaster and Pfeiffer, 2011, 2012). Other studies include the suppression of pathogenic lymphocytic choriomeningitis virus (LMCV) by co-infecting non-pathogenic strains (Teng and Oldstone, 1996); the suppression of high-fitness antigenic variants of FMDV clones that generated > 10^4-fold more progeny when removed from the surrounding low-fitness variants (Borrego *et al.*, 1993); or dominant interference by *trans*-acting variants of poliovirus that slowed the growth of drug-resistant mutants (Crowder and Kirkegaard, 2005). Further evidence for interference, a phenomenon weaker than suppression, was observed in viral populations transitioning towards extinction. In lethal mutagenesis studies of FMDV populations transitioning beyond the error threshold, viral RNA from the passage immediately preceding extinction interfered with infectious RNA under co-transfection (González-López *et al.*, 2004). This interference was not observed with unrelated 'junk' RNA or non-mutagenized, defective FMDV genomes. In this study, it was suggested that altered viral proteins resulting from the mutagenized RNA genomes might interfere with the replication of residual infectious genomes. Further co-transfections with mixtures of specific FMDV capsid or polymerase mutants together with the relevant FMDV RNA revealed that interference was observed only with variants that could still replicate (Perales *et al.*, 2007). In BHK-21 cells persistently infected with LCMV and treated with RNA mutagens, the accumulation of replication competent but non-infectious RNA prior to extinction also preceded the loss of replication competent RNA (Grande-Pérez *et al.*, 2005b). Termed 'defectors', these pre-extinction variants helped support *in silico* studies used to develop the lethal defection model of virus extinction (Iranzo and Manrubia, 2008).

Group contribution of minority variants to phenotype

One of the most defining aspects of considering RNA virus populations as quasispecies is the notion that phenotype is not simply the product of the consensus or master sequence, but is influenced by the presence of minority variants within the population. The evidence has often been anecdotal, such as observations that primary isolates of viruses and the cDNA infectious clones derived from the samples with the same consensus sequence, may have similar yet not perfectly matched growth characteristics. Although rarely formally reported in the literature, these observations suggest that something within these populations is different. The clearest demonstrations of this phenomenon are provided by studies of lethal mutagenesis, where mutagen-treated populations with severely debilitated fitness phenotypes present no differences in consensus sequence, while differences in the composition of minority genomes is readily identified (Grande-Pérez *et al.*, 2005a). Similarly, the studies on high- and low-fidelity variants of the Enteroviruses (poliovirus, CVB3 and EV71) all demonstrated attenuated phenotypes *in vivo*, despite no consensus sequence changes with respect to their wildtype virus counterparts (Gnädig *et al.*, 2012; Korboukh *et al.*, 2014; Meng and Kwang, 2014; Vignuzzi *et al.*, 2006). In a notable study on FMDV in mice, virus derived from the pancreata of mice was shown to be completely attenuated when injected into naïve mice; while virus from sera of these same mice retained virulence, with no differences in consensus sequence between samples (Sanz-Ramos *et al.*, 2008). However, identifying the specific minority variants among the thousands of mutants in the population that are responsible for these phenotypic differences remains a major challenge in the field. A recent study illustrates how this challenge may be tackled with the aid of deep sequencing. Bordería and colleagues monitored the evolution of coxsackievirus B3 in two cell lines during 40 passages, by whole-genome deep sequencing. They identified a handful of mutants that emerged over time, but were retained at subconsensus frequencies (ranging from 1% to 50% of the total population). No single genotype was able to match the phenotype manifested by the population; instead, they demonstrated that only a reconstituted quasispecies, composed of a mixture of the four most represented minority variants, conferred a fitness profile comparable to the original sample (Bordería *et al.*, 2015).

The genomics era: challenges and prospects of high-throughput sequencing technologies

The genomic age for eukaryotic virology was opened when the first complete genome sequence of one *Enterovirus*, poliovirus type 1 (Mahoney strain), was published in 1981 (Kitamura *et al.*, 1981; Racaniello and Baltimore, 1981a). This accomplishment allowed the direct genetic mapping of viral functions and lead to the development of reverse genetic systems (Racaniello and Baltimore, 1981b; Semler *et al.*, 1984; van der Werf *et al.*, 1986).

Over the last decade, the development of high-throughput sequencing (HTS) or next-generation sequencing technologies (NGS) has allowed a massive increase in capacity to sequence genomes. Although reads are generally shorter than by Sanger sequencing, NGS permits massive parallel sequencing of many genomes from individual templates (Metzker, 2010), and has redefined the *modus operandi* in virus genetics research by allowing the unprecedented generation of very large datasets on short time scales and affordable costs. This technology has already a tremendous impact in virology, although mostly centred around virus discovery (Briese *et al.*, 2014; Coffey *et al.*, 2014; Dereure *et al.*, 2013; Li *et al.*, 2013; Lipkin and Anthony, 2015; Mishra *et al.*, 2014; Ng *et al.*, 2013; Vasilakis *et al.*, 2013) and identifying the virome or microbiome, the collection of microbes present within a host (Delwart, 2013; Foulongne *et al.*, 2012; Hurwitz and Sullivan, 2013; MacLean *et al.*, 2009; Medini *et al.*, 2008; Tokarz *et al.*, 2014). Regarding Enteroviruses, metagenomic analysis has revealed a rich and diverse community of Enteroviruses in stool samples from children in South Asia (Kapoor *et al.*, 2008; Victoria *et al.*, 2009). A novel enterovirus 109 (HEV109) was discovered after deep sequencing nose and throat swab samples from Nicaraguan children with viral respiratory illness (Yozwiak *et al.*, 2010) and a novel enterovirus was identified in wild boars in Hungary (Boros *et al.*, 2012).

In regards to viral quasispecies analysis, the potential benefits of simultaneous sequencing millions of individual target sequences are clear. As a necessary step, differentiating sequencing error from true mutation, and adapting existing or creating new bioinformatics pipelines for these purposes is a current focus and main obstacle to future work (Beerenwinkel *et al.*, 2012). Nevertheless, the characterization of intra-host and intra-strain virus diversity and population dynamics for a variety of viruses has already begun. For example, the dynamics of emergence of immunity escape variants, antiviral resistance or receptor tropism has been examined for HIV and HCV (Archer *et al.*, 2012; Bull *et al.*, 2011; Love *et al.*, 2010), influenza virus (Selleri *et al.*, 2013) and plant viruses (Martínez *et al.*, 2012). Whole-genome analysis has also been used to examine overall changes in the genetic diversity of the virus population (Grad *et al.*, 2014; Töpfer *et al.*, 2013; Wright *et al.*, 2011). In a study involving experimental infection of humans with rhinovirus and whole-genome deep sequencing, the authors identified mutation 'hot-spots' across the capsid, 2C and 3C genes, and conserved regions in much of the non-structural genes (Cordey *et al.*, 2010). Currently, NGS approaches can readily detect minority variants at frequencies above of 1% of the reads, while bioinformatics and statistical treatments can drop this threshold to 0.1% or even further (Gregori *et al.*, 2014; Isakov *et al.*, 2015; Wilm *et al.*, 2012; Yang *et al.*, 2013b).

An alternative and elegant methodology to removing background error was developed by Acevedo and colleagues, called Circular resequencing (CirSeq) (Acevedo and Andino, 2014; Acevedo *et al.*, 2014) and was mostly developed using poliovirus as the model RNA population. This approach is based on the circularization of fragmented viral RNAs, which are then redundantly encoded into tandem repeats by 'rolling-circle' reverse transcription. When sequenced, the redundant copies within each read are aligned to derive a consensus sequence of their initial RNA template. A true mutation must therefore be read multiple times along the tandem repeats.

Another issue inherent to NGS is that the short read lengths make it difficult to reconstruct haplotypes, to determine which mutations co-localize on the same genotype. Ultimately, a representative construction of sequence space will require information on how many mutational steps one genome is with respects to another. In order to build up the genetic architecture of a viral population some approaches for quasispecies assembly have incorporated algorithms to determine the minimal set of haplotypes in a population (Eriksson *et al.*, 2008; Jojic *et al.*, 2008; Prosperi and Salemi, 2012; Zagordi *et al.*, 2010, 2011).

Final conclusions

High mutation rates and quasispecies dynamics confer great evolvability to RNA viruses and represent one of the major obstacles for the control and prevention of RNA viral diseases. The Enteroviruses, are no exception – in fact, they can be considered prime examples, with which to address key questions in virus evolution and population dynamics. With the increasing throughput and precision of sequencing technologies and the ever-growing precision of molecular studies, modelling the population dynamics of these RNA virus infections in real time seems to be within reach. In the coming years, the task of surmounting the current technical and conceptual hurdles will require biologists to design experiments in concert with computational scientists and applied mathematicians. Once new technologies and models are adequately optimized, probing the mutant swarm in a temporal and spatial fashion during infection should allow us to strip away the 'phenomenal' and reveal the underlying mechanism – to uncover the identities of the genotypes within *Enterovirus* populations that are playing key roles in adaptation, pathogenesis and transmission.

Glossary

Bottleneck In genetics, a bottleneck effect leads to a dramatic reduction in the number of individuals that can reproduce. Bottlenecks stochastically reduce genetic variation and are not necessarily selective events.

Complementation A process by which a defective virus can take advantage of functional products (nucleic acid sequences or proteins) from another virus that is infecting the same cell. As a result, the defective virus does not experience loss of fitness from its mutation (or mutations).

Consensus (or average) sequence The sequence presenting the most frequent residue (nucleotide or amino acid) found at each position within the entire population. The consensus sequence may not physically exist within the **mutant spectrum**.

Digital organisms Self-replicating computer programs that mutate and evolve, often in competition with each other for CPU (central processing unit) cycles.

Epistasis Interactions between mutations such that their combined effect on fitness is different to that expected from their effects in isolation.

Error catastrophe The loss of meaningful genetic information when a population is pushed beyond its maximum mutation rate. In theoretical models, the error catastrophe has been compared to a chemical phase transition, and a true error catastrophe has not been observed experimentally.

Evolvability The capacity of an organism with a particular genotype to generate adaptive genetic diversity, also termed evolutionary adaptability.
Fitness landscapes Topographical models that link fitness values to individual sequences.
Fitness The ability of an entity to survive and reproduce in a given environment. In experimental virology, replicative efficiency is often used as a surrogate for fitness. It is defined as the capacity of a virus to generate infectious progeny in a specific environment. Fitness is a relative measure, often with respect to either a 'wildtype' or average genotype.
Hamming distance Number of mutated positions from the best adapted genotype in a mutant distribution. It is also used to mean the number of nucleotide differences between two sequences.
Lethal mutagenesis The process whereby the number of viable individual viruses in a population is reduced through increases in the mutation rate and the accumulation of lethal mutations.
Master sequence The dominant genomic nucleotide sequence in a quasispecies, generally corresponding to the highest fitness in that given environment. It may or may not coincide with the **consensus sequence.** The master sequence can change as the environment is modified.
Mutant spectrum The ensemble of mutant genomes that compose a **quasispecies.** It is also referred to as *mutant swarm* or *cloud.*
Mutation frequency The proportion of mutated sites in a population of genomes. It can be calculated for an entire sequence or for a specific site of a genome (e.g. in the frequency of monoclonal antibody-escape mutants), for all mutation types, or for a specific mutation type (nucleotide substitutions, transition/transversions, insertions/deletions).
Mutation rate The rate of occurrence of a mutational event during genome replication. Unlike mutation frequency, mutation rate implies a unit of time e.g. mutations per replication cycle. In the literature of population genetics, mutation rate is often used to mean the evolutionary rate (fixation or accumulation of mutations).
Mutational fitness effect The effects of mutations on fitness; often described in a model that combines both the strength and distribution of these effects.
Mutator/antimutator Originally described in bacteria, a mutator generates more mutations than 'wildtype' and an antimutator generates less. Whereas low- and high-fidelity variants generally refer to the polymerase enzyme, mutators/antimutators are a broader term whose changes in mutation rates may also include non-polymerase fidelity mechanisms.
Negative selection The removal of deleterious alleles from a population by natural selection. Also called **purifying selection**.
Phylodynamic study A study that develops a quantitative model, incorporating both a pathogen phylogeny and epidemiological or immunological data, to describe an infectious disease.
Sequence space All possible permutations of a given nucleic acid or amino acid sequence. Sequence space can be mathematically described as a *n*-dimensional hypercube occupied by up to 4n (nucleic acid) or 20n (protein) distinct sequences or coordinates.
Synonymous mutations Codon mutations that do not alter the amino acid specificity of the codons. By contrast, non-synonymous mutations change the encoded amino acid.

References

Aaskov, J., Buzacott, K., Thu, H.M., Lowry, K., and Holmes, E.C. (2006). Long-term transmission of defective RNA viruses in humans and Aedes mosquitoes. Science *311*, 236–238.

Acevedo, A., and Andino, R. (2014). Library preparation for highly accurate population sequencing of RNA viruses. Nat. Protoc. *9*, 1760–1769.

Acevedo, A., Brodsky, L., and Andino, R. (2014). Mutational and fitness landscapes of an RNA virus revealed through population sequencing. Nature *505*, 686–690.

Ali, A., Li, H., Schneider, W.L., Sherman, D.J., Gray, S., Smith, D., and Roossinck, M.J. (2006). Analysis of genetic bottlenecks during horizontal transmission of Cucumber mosaic virus. J. Virol. *80*, 8345–8350.

Archer, J., Weber, J., Henry, K., Winner, D., Gibson, R., Lee, L., Paxinos, E., Arts, E.J., Robertson, D.L., Mimms, L., and Quiñones-Mateu, M.E. (2012). Use of four next-generation sequencing platforms to determine HIV-1 coreceptor tropism. PLoS One 7, e49602.

Arias, A., Arnold, J.J., Sierra, M., Smidansky, E.D., Domingo, E., and Cameron, C.E. (2008). Determinants of RNA-dependent RNA polymerase (in)fidelity revealed by kinetic analysis of the polymerase encoded by a foot-and-mouth disease virus mutant with reduced sensitivity to ribavirin. J. Virol. *82*, 12346–12355.

Arnold, J.J., Vignuzzi, M., Stone, J.K., Andino, R., and Cameron, C.E. (2005). Remote site control of an active site fidelity checkpoint in a viral RNA-dependent RNA polymerase. J. Biol. Chem. *280*, 25706–25716.

Azevedo, R.B., Lohaus, R., Srinivasan, S., Dang, K.K., and Burch, C.L. (2006). Sexual reproduction selects for robustness and negative epistasis in artificial gene networks. Nature *440*, 87–90.

Batschelet, E., Domingo, E., and Weissmann, C. (1976). The proportion of revertant and mutant phage in a growing population, as a function of mutation and growth rate. Gene *1*, 27–32.

Beaucourt, S., Bordería, A.V., Coffey, L.L., Gnädig, N.F., Sanz-Ramos, M., Beeharry, Y., and Vignuzzi, M. (2011). Isolation of fidelity variants of RNA viruses and characterization of virus mutation frequency. J. Vis. Exp. *52*, 2953.

Beaucourt, S., and Vignuzzi, M. (2014). Ribavirin: a drug active against many viruses with multiple effects on virus replication and propagation. Molecular basis of ribavirin resistance. Curr. Opin. Virol. *8*, 10–15.

Beerenwinkel, N., Günthard, H.F., Roth, V., and Metzner, K.J. (2012). Challenges and opportunities in estimating viral genetic diversity from next-generation sequencing data. Front. Microbiol. *3*, 329.

Belshaw, R., Sanjuán, R., and Pybus, O.G. (2011). Viral mutation and substitution: units and levels. Curr. Opin. Virol. *1*, 430–435.

Betancourt, M., Fereres, A., Fraile, A., and García-Arenal, F. (2008). Estimation of the effective number of founders that initiate an infection after aphid transmission of a multipartite plant virus. J. Virol. *82*, 12416–12421.

Bible, J.M., Iturriza-Gomara, M., Megson, B., Brown, D., Pantelidis, P., Earl, P., Bendig, J., and Tong, C.Y. (2008). Molecular epidemiology of human enterovirus 71 in the United Kingdom from 1998 to 2006. J. Clin. Microbiol. *46*, 3192–3200.

Bordería, A.V., Isakov, O., Moratorio, G., Henningsson, R., Aguera-Gonzalez, S., Organtini, L., Gnädig, N.F., Blanc, H., Alcover, A., Hafenstein, S., Fontes, M., Shomron, N., and Vignuzzi, M.. (2015). Group Selection and Contribution of Minority Variants during Virus Adaptation Determines Virus Fitness and Phenotype. PLoS Pathog. *11*, e1004838.

Boros, Á., Nemes, C., Pankovics, P., Bíró, H., Kapusinszky, B., Delwart, E., and Reuter, G. (2012). Characterization of a novel porcine enterovirus in wild boars in Hungary. Arch. Virol. *157*, 981–986.

Borrego, B., Novella, I.S., Giralt, E., Andreu, D., and Domingo, E. (1993). Distinct repertoire of antigenic variants of foot-and-mouth disease virus in the presence or absence of immune selection. J. Virol. *67*, 6071–6079.

Bouslama, L., Nasri, D., Chollet, L., Belguith, K., Bourlet, T., Aouni, M., Pozzetto, B., and Pillet, S. (2007). Natural recombination event within the capsid genomic region leading to a chimeric strain of human enterovirus B. J. Virol. *81*, 8944–8952.

Briese, T., Chowdhary, R., Travassos da Rosa, A., Hutchison, S.K., Popov, V., Street, C., Tesh, R.B., and Lipkin, W.I. (2014). Upolu virus and Aransas Bay virus, two presumptive bunyaviruses, are novel members of the family Orthomyxoviridae. J. Virol. *88*, 5298–5309.

Brown, B., Oberste, M.S., Maher, K., and Pallansch, M.A. (2003). Complete genomic sequencing shows that polioviruses and members of human enterovirus species C are closely related in the noncapsid coding region. J. Virol. 77, 8973–8984.

Bull, R.A., Luciani, F., McElroy, K., Gaudieri, S., Pham, S.T., Chopra, A., Cameron, B., Maher, L., Dore, G.J., White, P.A., and Lloyd, A.R. (2011). Sequential bottlenecks drive viral evolution in early acute hepatitis C virus infection. PLoS Pathog. 7, e1002243.

Campagnola, G., McDonald, S., Beaucourt, S., Vignuzzi, M., and Peersen, O.B. (2015). Structure-function relationships underlying the replication fidelity of viral RNA-dependent RNA polymerases. J. Virol. *89*, 275-286.

Chumakov, K.M., Powers, L.B., Noonan, K.E., Roninson, I.B., and Levenbook, I.S. (1991). Correlation between amount of virus with altered nucleotide-sequence and the monkey test for acceptability of oral poliovirus vaccine. Proc. Natl. Acad. Sci. U.S.A. *88*, 199–203.

Clarke, D.K., Duarte, E.A., Moya, A., Elena, S.F., Domingo, E., and Holland, J. (1993). Genetic bottlenecks and population passages cause profound fitness differences in RNA viruses. J. Virol. *67*, 222–228.

Clarke, D.K., Duarte, E.A., Elena, S.F., Moya, A., Domingo, E., and Holland, J. (1994). The red queen reigns in the kingdom of RNA viruses. Proc. Natl. Acad. Sci. U.S.A. *91*, 4821–4824.

Codoñer, F.M., Darós, J.A., Solé, R.V., and Elena, S.F. (2006). The fittest versus the flattest: experimental confirmation of the quasispecies effect with subviral pathogens. PLoS Pathog. *2*, e136.

Coffey, L.L., Page, B.L., Greninger, A.L., Herring, B.L., Russell, R.C., Doggett, S.L., Haniotis, J., Wang, C., Deng, X., and Delwart, E.L. (2014). Enhanced arbovirus surveillance with deep sequencing:

identification of novel rhabdoviruses and bunyaviruses in Australian mosquitoes. Virology *448*, 146–158.
Combelas, N., Holmblat, B., Joffret, M.L., Colbère-Garapin, F., and Delpeyroux, F. (2011). Recombination between poliovirus and coxsackie A viruses of species C: a model of viral genetic plasticity and emergence. Viruses *3*, 1460–1484.
Copper, P.D., Steiner-Pryor, A., Scotti, P.D., and Delong, D. (1974). On the nature of poliovirus genetic recombinants. J. Gen. Virol. *23*, 41–49.
Cordey, S., Junier, T., Gerlach, D., Gobbini, F., Farinelli, L., Zdobnov, E.M., Winther, B., Tapparel, C., and Kaiser, L. (2010). Rhinovirus genome evolution during experimental human infection. PLoS One *5*, e10588.
Crotty, S., Maag, D., Arnold, J.J., Zhong, W., Lau, J.Y., Hong, Z., Andino, R., and Cameron, C.E. (2000). The broad-spectrum antiviral ribonucleoside ribavirin is an RNA virus mutagen. Nat. Med. *6*, 1375–1379.
Crotty, S., Cameron, C.E., and Andino, R. (2001). RNA virus error catastrophe: direct molecular test by using ribavirin. Proc. Natl. Acad. Sci. U.S.A. *98*, 6895–6900.
Crotty, S., Cameron, C., and Andino, R. (2002). Ribavirin's antiviral mechanism of action: lethal mutagenesis? J. Mol. Med. *80*, 86–95.
Crowder, S., and Kirkegaard, K. (2005). Trans-dominant inhibition of RNA viral replication can slow growth of drug-resistant viruses. Nat. Genet. *37*, 701–709.
da Silva, J., Coetzer, M., Nedellec, R., Pastore, C., and Mosier, D.E. (2010). Fitness epistasis and constraints on adaptation in a human immunodeficiency virus type 1 protein region. Genetics *185*, 293–303.
de la Iglesia, F., and Elena, S.F. (2007). Fitness declines in Tobacco etch virus upon serial bottleneck transfers. J. Virol. *81*, 4941-4947.
de la Torre, J.C., and Holland, J.J. (1990). RNA virus quasispecies populations can suppress vastly superior mutant progeny. J. Virol. *64*, 6278-6281.
de la Torre, J.C., Wimmer, E., and Holland, J.J. (1990). Very high frequency of reversion to guanidine resistance in clonal pools of guanidine-dependent type 1 poliovirus. J. Virol. *64*, 664–671.
de la Torre, J.C., Giachetti, C., Semler, B.L., and Holland, J.J. (1992). High frequency of single-base transitions and extreme frequency of precise multiple-base reversion mutations in poliovirus. Proc. Natl. Acad. Sci. U.S.A. *89*, 2531–2535.
de Visser, J.A., and Krug, J. (2014). Empirical fitness landscapes and the predictability of evolution. Nat. Rev. Genet. *15*, 480–490.
Delwart, E. (2013). A roadmap to the human virome. PLoS Pathog. *9*, e1003146.
Dereure, O., Cheval, J., Du Thanh, A., Pariente, K., Sauvage, V., Manuguerra, J.C., Caro, V., Foulongne, V., and Eloit, M. (2013). No evidence for viral sequences in mycosis fungoides and Sézary syndrome skin lesions: a high-throughput sequencing approach. J. Invest. Dermatol. *133*, 853–855.
Domingo, E., Sabo, D., Taniguchi, T., and Weissmann, C. (1978). Nucleotide sequence heterogeneity of an RNA phage population. Cell *13*, 735–744.
Domingo, E., Escarmis, C., Martinez, M.A., Martinez-Salas, E., and Mateu, M.G. (1992). Foot-and-mouth disease virus populations are quasispecies. Curr. Top. Microbiol. Immunol. *176*, 33–47.
Domingo, E., Holland, J.J., Biebricher, C., and Eigen, M. (1995). Quasispecies: the concept and the word. In Molecular evolution of the viruses, A. Gibbs, C. Calisher and F. García-Arenal, eds. (Cambridge, UK: Cambridge University Press), pp. 171-180.
Domingo, E., and Holland, J.J. (1997). RNA virus mutations and fitness for survival. Annu. Rev. Microbiol. *51*, 151–178.
Domingo, E. (2006). Quasispecies: concept and implications for virology (Heidelberg, Germany: Springer).
Domingo-Calap, P., Cuevas, J.M., and Sanjuán, R. (2009). The fitness effects of random mutations in single-stranded DNA and RNA bacteriophages. PLoS Genet. *5*, e1000742.
Drake, J.W., and Holland, J.J. (1999). Mutation rates among RNA viruses. Proc. Natl. Acad. Sci. U.S.A. *96*, 13910–13913.
Duarte, E., Clarke, D., Moya, A., Domingo, E., and Holland, J. (1992). Rapid fitness losses in mammalian RNA virus clones due to Muller's ratchet. Proc. Natl. Acad. Sci. U.S.A. *89*, 6015–6019.
Duarte, E.A., Novella, I.S., Ledesma, S., Clarke, D.K., Moya, A., Elena, S.F., Domingo, E., and Holland, J.J. (1994). Subclonal components of consensus fitness in an RNA virus clone. J. Virol. *68*, 4295–4301.
Eggers, H.J., and Tamm, I. (1965). Coxsackie A9 virus: mutation from drug dependence to drug independence. Science *148*, 97–98.
Eigen, M. (1971). Selforganization of matter and the evolution of biological macromolecules. Naturwissenschaften *58*, 465–523.

Eigen, M., and Schuster, P. (1977). The hypercycle. A principle of natural self-organization. Part A: Emergence of the hypercycle. Naturwissenschaften *64*, 541–565.
Eigen, M., and Biebricher, C.K. (1988). Sequence space and quasispecies distribution. In RNA Genetics, E. Domingo, J.J. Holland, and R. Ahlquist, eds. (Boca Raton, FL: CRC Press), pp. 211–245.
Eigen, M., McCaskill, J., and Schuster, P. (1988). Molecular quasi-species. J. Phys. Chem. *92*, 6881–6891.
Eigen, M. (1992). Steps towards life: A perspective on evolution (Oxford, UK: Oxford University Press).
Elena, S.F., and Sanjuán, R. (2005). Adaptive value of high mutation rates of RNA viruses: separating causes from consequences. J. Virol. *79*, 11555–11558.
Elena, S.F., and Sanjuán, R. (2008). The effect of genetic robustness on evolvability in digital organisms. BMC Evol. Biol. *8*, 284.
Elena, S.F., Solé, R.V., and Sardanyés, J. (2010). Simple genomes, complex interactions: epistasis in RNA virus. Chaos *20*, 026106.
Elena, S.F. (2012). RNA virus genetic robustness: possible causes and some consequences. Curr. Opin. Virol. *2*, 525–530.
Epstein, I.R., and Eigen, M. (1979). Selection and self-organization of self-reproducing macromolecules under the constraint of constant flux. Biophys. Chem. *10*, 153–160.
Eriksson, N., Pachter, L., Mitsuya, Y., Rhee, S.Y., Wang, C., Gharizadeh, B., Ronaghi, M., Shafer, R.W., and Beerenwinkel, N. (2008). Viral population estimation using pyrosequencing. PLoS Comput. Biol. *4*, e1000074.
Escarmís, C., Dávila, M., Charpentier, N., Bracho, A., Moya, A., and Domingo, E. (1996). Genetic lesions associated with Muller's ratchet in an RNA virus. J. Mol. Biol. *264*, 255–267.
Escarmís, C., Dávila, M., and Domingo, E. (1999). Multiple molecular pathways for fitness recovery of an RNA virus debilitated by operation of Muller's ratchet. J. Mol. Biol. *285*, 495–505.
Esteves-Jaramillo, A., Estívariz, C.F., Peñaranda, S., Richardson, V.L., Reyna, J., Coronel, D.L., Carrión, V., Landaverde, J.M., Wassilak, S.G., Pérez-Sánchez, E.E., López-Martínez, I., Burns, C.C., and Pallansch, M.A. (2014). Detection of vaccine-derived polioviruses in Mexico using environmental surveillance. J. Infect. Dis. *210*, 315–323.
Ferrer-Orta, C., Agudo, R., Domingo, E., and Verdaguer, N. (2009). Structural insights into replication initiation and elongation processes by the FMDV RNA-dependent RNA polymerase. Curr. Opin. Struct. Biol. *19*, 752–758.
Foulongne, V., Sauvage, V., Hebert, C., Dereure, O., Cheval, J., Gouilh, M.A., Pariente, K., Segondy, M., Burguière, A., Manuguerra, J.C., Caro, V., and Eloit, M. (2012). Human skin microbiota: high diversity of DNA viruses identified on the human skin by high throughput sequencing. PLoS One *7*, e38499.
García-Arriaza, J., Manrubia, S.C., Toja, M., Domingo, E., and Escarmís, C. (2004). Evolutionary transition toward defective RNAs that are infectious by complementation. J. Virol. *78*, 11678–11685.
García-Arriaza, J., Ojosnegros, S., Dávila, M., Domingo, E., and Escarmís, C. (2006). Dynamics of mutation and recombination in a replicating population of complementing, defective viral genomes. J. Mol. Biol. *360*, 558–572.
Gelderblom, H.C., Vatakis, D.N., Burke, S.A., Lawrie, S.D., Bristol, G.C., and Levy, D.N. (2008). Viral complementation allows HIV-1 replication without integration. Retrovirology *5*, 60.
Gnädig, N.F., Beaucourt, S., Campagnola, G., Bordería, A.V., Sanz-Ramos, M., Gong, P., Blanc, H., Peersen, O.B., and Vignuzzi, M. (2012). Coxsackievirus B3 mutator strains are attenuated in vivo. Proc. Natl. Acad. Sci. U.S.A. *109*, E2294–303.
Goldhill, D., Lee, A., Williams, E.S., and Turner, P.E. (2014). Evolvability and robustness in populations of RNA virus Φ6. Front. Microbiol. *5*, 35.
Gong, P., and Peersen, O.B. (2010). Structural basis for active site closure by the poliovirus RNA-dependent RNA polymerase. Proc. Natl. Acad. Sci. U.S.A. *107*, 22505–22510.
González-López, C., Arias, A., Pariente, N., Gómez-Mariano, G., and Domingo, E. (2004). Preextinction viral RNA can interfere with infectivity. J. Virol. *78*, 3319–3324.
Graci, J.D., Gnädig, N.F., Galarraga, J.E., Castro, C., Vignuzzi, M., and Cameron, C.E. (2012). Mutational robustness of an RNA virus influences sensitivity to lethal mutagenesis. J. Virol. *86*, 2869–2873.
Grad, Y.H., Newman, R., Zody, M., Yang, X., Murphy, R., Qu, J., Malboeuf, C.M., Levin, J.Z., Lipsitch, M., and DeVincenzo, J. (2014). Within-host whole-genome deep sequencing and diversity analysis of human respiratory syncytial virus infection reveals dynamics of genomic diversity in the absence and presence of immune pressure. J. Virol. *88*, 7286–7293.
Grande-Pérez, A., Gómez-Mariano, G., Lowenstein, P.R., and Domingo, E. (2005a). Mutagenesis-induced, large fitness variations with an invariant arenavirus consensus genomic nucleotide sequence. J. Virol. *79*, 10451-10459.

Grande-Pérez, A., Lázaro, E., Lowenstein, P., Domingo, E., and Manrubia, S.C. (2005b). Suppression of viral infectivity through lethal defection. Proc. Natl. Acad. Sci. U.S.A. *102*, 4448–4452.

Gregori, J., Salicrú, M., Domingo, E., Sanchez, A., Esteban, J.I., Rodríguez-Frías, F., and Quer, J. (2014). Inference with viral quasispecies diversity indices: clonal and NGS approaches. Bioinformatics *30*, 1104-1111..

Haaland, R.E., Hawkins, P.A., Salazar-Gonzalez, J., Johnson, S., Tichacek, A., Karita, E., Manigart, O., Mulenga, J., Keele, B.F., Shaw, G.M., Hahn, B.H., Allen, S.A., Derdeyn, C.A., and Hunter, E. (2009). Inflammatory genital infections mitigate a severe genetic bottleneck in heterosexual transmission of subtype A and C HIV-1. PLoS Pathog. *5*, e1000274.

Harvala, H., Sharp, C.P., Ngole, E.M., Delaporte, E., Peeters, M., and Simmonds, P. (2011). Detection and genetic characterization of enteroviruses circulating among wild populations of chimpanzees in Cameroon: relationship with human and simian enteroviruses. J. Virol. *85*, 4480–4486.

Hogeweg, P. (2012). Toward a theory of multilevel evolution: long-term information integration shapes the mutational landscape and enhances evolvability. Adv. Exp. Med. Biol. *751*, 195–224.

Holland, J.J. (2006). Transitions in understanding of RNA viruses: a historical perspective. Curr. Top. Microbiol. Immunol. *299*, 371–401.

Holmes, E.C. (2003). Error thresholds and the constraints to RNA virus evolution. Trends Microbiol. *11*, 543–546.

Holmes, E.C. (2010). The RNA virus quasispecies: fact or fiction? J. Mol. Biol. *400*, 271–273.

Holmes, E.C., and Moya, A. (2002). Is the quasispecies concept relevant to RNA viruses? J. Virol. *76*, 460–465.

Huang, T., Wang, W., Bessaud, M., Ren, P., Sheng, J., Yan, H., Zhang, J., Lin, X., Wang, Y., Delpeyroux, F., and Deubel, V. (2009). Evidence of recombination and genetic diversity in human rhinoviruses in children with acute respiratory infection. PLoS One *4*, e6355.

Hurwitz, B.L., and Sullivan, M.B. (2013). The Pacific Ocean virome (POV): a marine viral metagenomic dataset and associated protein clusters for quantitative viral ecology. PLoS One *8*, e57355.

Iranzo, J., and Manrubia, S.C. (2008). Stochastic extinction of viral infectivity through the action of defectors. EPL. *85*, 18001.

Iranzo, J., and Manrubia, S.C. (2012). Evolutionary dynamics of genome segmentation in multipartite viruses. Proc. Biol. Sci. *279*, 3812–3819.

Isakov, O., Bordería, A.V., Golan, D., Hamenahem, A., Celniker, G., Yoffe, L., Blanc, H., Vignuzzi, M., and Shomron, N. (2015). Deep sequencing analysis of viral infection and evolution allows rapid and detailed characterization of viral mutant spectrum. Bioinformatics *31*, 1–9.

Jegouic, S., Joffret, M.L., Blanchard, C., Riquet, F.B., Perret, C., Pelletier, I., Colbere-Garapin, F., Rakoto-Andrianarivelo, M., and Delpeyroux, F. (2009). Recombination between polioviruses and co-circulating Coxsackie A viruses: role in the emergence of pathogenic vaccine-derived polioviruses. PLoS Pathog. *5*, e1000412.

Jenkins, G.M., Worobey, M., Woelk, C.H., and Holmes, E.C. (2001). Evidence for the non-quasispecies evolution of RNA viruses. Mol. Biol. Evol. *18*, 987–994.

Jenkins, G.M., Rambaut, A., Pybus, O.G., and Holmes, E.C. (2002). Rates of molecular evolution in RNA viruses: a quantitative phylogenetic analysis. J. Mol. Evol. *54*, 156–165.

Jojic, V., Hertz, T., and Jojic, N. (2008). Population sequencing using short reads: HIV as a case study. Pac. Symp. Biocomput. 114–125.

Kapoor, A., Victoria, J., Simmonds, P., Slikas, E., Chieochansin, T., Naeem, A., Shaukat, S., Sharif, S., Alam, M.M., Angez, M., Wang, C., Shafer, R.W., Zaidi, S., and Delwart, E. (2008). A highly prevalent and genetically diversified Picornaviridae genus in South Asian children. Proc. Natl. Acad. Sci. U.S.A. *105*, 20482–20487.

Kitamura, N., Semler, B.L., Rothberg, P.G., Larsen, G.R., Adler, C.J., Dorner, A.J., Emini, E.A., Hanecak, R., Lee, J.J., van der Werf, S., Anderson, C.W., and Wimmer, E. (1981). Primary structure, gene organization and polypeptide expression of poliovirus RNA. Nature *291*, 547–553.

Korboukh, V.K., Lee, C.A., Acevedo, A., Vignuzzi, M., Xiao, Y., Arnold, J.J., Hemperly, S., Graci, J.D., August, A., Andino, R., and Cameron, C.E. (2014). RNA virus population diversity, an optimum for maximal fitness and virulence. J. Biol. Chem. *289*, 29531–29544.

Kuss, S.K., Etheredge, C.A., and Pfeiffer, J.K. (2008). Multiple host barriers restrict poliovirus trafficking in mice. PLoS Pathog. *4*, e1000082.

Lancaster, K.Z., and Pfeiffer, J.K. (2011). Mechanisms controlling virulence thresholds of mixed viral populations. J. Virol. *85*, 9778–9788.

Lancaster, K.Z., and Pfeiffer, J.K. (2012). Viral population dynamics and virulence thresholds. Curr. Opin. Microbiol. *15*, 525–530.
Lauring, A.S., and Andino, R. (2011). Exploring the fitness landscape of an RNA virus by using a universal barcode microarray. J. Virol. *85*, 3780–3791.
Lauring, A.S., Acevedo, A., Cooper, S.B., and Andino, R. (2012). Codon usage determines the mutational robustness, evolutionary capacity, and virulence of an RNA virus. Cell Host Microbe *12*, 623–632.
Le, Nouën, C., Brock, L.G., Luongo, C., McCarty, T., Yang, L., Mehedi, M., Wimmer, E., Mueller, S., Collins, P.L., Buchholz, U.J., and DiNapoli, J.M. (2014). Attenuation of human respiratory syncytial virus by genome-scale codon-pair deoptimization. Proc. Natl. Acad. Sci. U.S.A. *111*, 13169–13174.
Ledinko, N. (1963). Genetic recombination with poliovirus type 1. Studies of crosses between a normal horse serum-resistant mutant and several guanidine-resistant mutants of the same strain. Virology *20*, 107–119.
Levi, L.I., Gnädig, N.F., Beaucourt, S., McPherson, M.J., Baron, B., Arnold, J.J., and Vignuzzi, M. (2010). Fidelity variants of RNA dependent RNA polymerases uncover an indirect, mutagenic activity of amiloride compounds. PLoS Pathog. *6*, e1001163.
Li, H., and Roossinck, M.J. (2004). Genetic bottlenecks reduce population variation in an experimental RNA virus population. J. Virol. *78*, 10582–10587.
Li, L., Diab, S., McGraw, S., Barr, B., Traslavina, R., Higgins, R., Talbot, T., Blanchard, P., Rimoldi, G., Fahsbender, E., Page, B., Phan, T.G., Wang, C., Deng, X., Pesavento, P., and Delwart, E. (2013). Divergent astrovirus associated with neurologic disease in cattle. Emerg. Infect. Dis. *19*, 1385–1392.
Lipkin, W.I., and Anthony, S.J. (2015). Virus hunting. Virology 479-480, 194–199.
Liu, X., Yang, X., Lee, C.A., Moustafa, I.M., Smidansky, E.D., Lum, D., Arnold, J.J., Cameron, C.E., and Boehr, D.D. (2013). Vaccine-derived mutation in motif D of poliovirus RNA-dependent RNA polymerase lowers nucleotide incorporation fidelity. J. Biol. Chem. *288*, 32753–32765.
Lorenzo-Redondo, R., Delgado, S., Morán, F., and Lopez-Galindez, C. (2014). Realistic three dimensional fitness landscapes generated by self organizing maps for the analysis of experimental HIV-1 evolution. PLoS One *9*, e88579.
Love, T.M., Thurston, S.W., Keefer, M.C., Dewhurst, S., and Lee, H.Y. (2010). Mathematical modeling of ultradeep sequencing data reveals that acute CD8+ T-lymphocyte responses exert strong selective pressure in simian immunodeficiency virus-infected macaques but still fail to clear founder epitope sequences. J. Virol. *84*, 5802–5814.
Lukashev, A.N. (2010). Recombination among picornaviruses. Rev. Med. Virol. *20*, 327–337.
Lukashev, A.N., Lashkevich, V.A., Ivanova, O.E., Koroleva, G.A., Hinkkanen, A.E., and Ilonen, J. (2005). Recombination in circulating Human enterovirus B: independent evolution of structural and non-structural genome regions. J. Gen. Virol. *86*, 3281–3290.
Lukashev, A.N., Shumilina, E.Y., Belalov, I.S., Ivanova, O.E., Eremeeva, T.P., Reznik, V.I., Trotsenko, O.E., Drexler, J.F., and Drosten, C. (2014). Recombination strategies and evolutionary dynamics of the Human enterovirus A global gene pool. J. Gen. Virol. *95*, 868–873.
MacLean, D., Jones, J.D., and Studholme, D.J. (2009). Application of 'next-generation' sequencing technologies to microbial genetics. Nat. Rev. Microbiol. *7*, 287–296.
Martínez, M.A., Carrillo, C., González-Candelas, F., Moya, A., Domingo, E., and Sobrino, F. (1991). Fitness alteration of foot-and-mouth disease virus mutants: measurement of adaptability of viral quasispecies. J. Virol. *65*, 3954–3957.
Martínez, F., Lafforgue, G., Morelli, M.J., González-Candelas, F., Chua, N.H., Daròs, J.A., and Elena, S.F. (2012). Ultradeep sequencing analysis of population dynamics of virus escape mutants in RNAi-mediated resistant plants. Mol. Biol. Evol. *29*, 3297–3307.
Masel, J., and Siegal, M.L. (2009). Robustness: mechanisms and consequences. Trends Genet. *25*, 395–403.
McIntyre, C.L., McWilliam Leitch, E.C., Savolainen-Kopra, C., Hovi, T., and Simmonds, P. (2010). Analysis of genetic diversity and sites of recombination in human rhinovirus species C. J. Virol. *84*, 10297–10310.
McWilliam Leitch, E.C., Bendig, J., Cabrerizo, M., Cardosa, J., Hyypiä, T., Ivanova, O.E., Kelly, A., Kroes, A.C., Lukashev, A., MacAdam, A., McMinn, P., Roivainen, M., Trallero, G., Evans, D.J., and Simmonds, P. (2009). Transmission networks and population turnover of echovirus 30. J. Virol. *83*, 2109–2118.
McWilliam Leitch, E.C., Cabrerizo, M., Cardosa, J., Harvala, H., Ivanova, O.E., Kroes, A.C., Lukashev, A., Muir, P., Odoom, J., Roivainen, M., Susi, P., Trallero, G., Evans, D.J., and Simmonds, P. (2010). Evolutionary dynamics and temporal/geographical correlates of recombination in the human enterovirus echovirus types 9, 11, and 30. J. Virol. *84*, 9292–9300.

McWilliam Leitch, E.C., Cabrerizo, M., Cardosa, J., Harvala, H., Ivanova, O.E., Koike, S., Kroes, A.C., Lukashev, A., Perera, D., Roivainen, M., Susi, P., Trallero, G., Evans, D.J., and Simmonds, P. (2012). The association of recombination events in the founding and emergence of subgenogroup evolutionary lineages of human enterovirus 71. J. Virol. *86*, 2676–2685.

Medini, D., Serruto, D., Parkhill, J., Relman, D.A., Donati, C., Moxon, R., Falkow, S., and Rappuoli, R. (2008). Microbiology in the post-genomic era. Nat. Rev. Microbiol. *6*, 419–430.

Meng, T., and Kwang, J. (2014). Attenuation of human enterovirus 71 high-replication-fidelity variants in AG129 mice. J. Virol. *88*, 5803–5815.

Metzker, M.L. (2010). Sequencing technologies - the next generation. Nat. Rev. Genet. *11*, 31–46.

Mirand, A., Henquell, C., Archimbaud, C., Peigue-Lafeuille, H., and Bailly, J.L. (2007). Emergence of recent echovirus 30 lineages is marked by serial genetic recombination events. J. Gen. Virol. *88*, 166–176.

Mishra, N., Pereira, M., Rhodes, R.H., An, P., Pipas, J.M., Jain, K., Kapoor, A., Briese, T., Faust, P.L., and Lipkin, W.I. (2014). Identification of a novel polyomavirus in a pancreatic transplant recipient with retinal blindness and vasculitic myopathy. J. Infect. Dis. *210*, 1595–1599.

Moratorio, G., Henningsson, R., Barbezange, C., Carrau, L., Bordería, A.V., Blanc, H., Beaucourt, S., Poirier, E.Z., Vallet, T., Boussier, J., Mounce, B.C., Fontes, M., and Vignuzzi, M. (2017). Attenuation of RNA viruses by redirecting their evolution in sequence space. Nat. Microbiol. *2*, 17088.

Moreno, I.M., Malpica, J.M., Rodríguez-Cerezo, E., and García-Arenal, F. (1997). A mutation in tomato aspermy cucumovirus that abolishes cell-to-cell movement is maintained to high levels in the viral RNA population by complementation. J. Virol. *71*, 9157–9162.

Ng, T.F., Driscoll, C., Carlos, M.P., Prioleau, A., Schmieder, R., Dwivedi, B., Wong, J., Cha, Y., Head, S., Breitbart, M., and Delwart, E. (2013). Distinct lineage of vesiculovirus from big brown bats, United States. Emerg. Infect. Dis. *19*, 1978–1980.

Novella, I.S., Duarte, E.A., Elena, S.F., Moya, A., Domingo, E., and Holland, J.J. (1995a). Exponential increases of RNA virus fitness during large population transmissions. Proc. Natl. Acad. Sci. U.S.A. *92*, 5841-5844.

Novella, I.S., Elena, S.F., Moya, A., Domingo, E., and Holland, J.J. (1995b). Size of genetic bottlenecks leading to virus fitness loss is determined by mean initial population fitness. J. Virol. *69*, 2869–2872.

Novella, I.S., Quer, J., Domingo, E., and Holland, J.J. (1999). Exponential fitness gains of RNA virus populations are limited by bottleneck effects. J. Virol. *73*, 1668–1671.

Novella, I.S. (2004). Negative effect of genetic bottlenecks on the adaptability of vesicular stomatitis virus. J. Mol. Biol. *336*, 61–67.

Novella, I.S., and Ebendick-Corpus, B.E. (2004). Molecular basis of fitness loss and fitness recovery in vesicular stomatitis virus. J. Mol. Biol. *342*, 1423–1430.

Novella, I.S., Reissig, D.D., and Wilke, C.O. (2004). Density-dependent selection in vesicular stomatitis virus. J. Virol. *78*, 5799–5804.

Novella, I.S., Dutta, R.N., and Wilke, C.O. (2008). A linear relationship between fitness and the logarithm of the critical bottleneck size in vesicular stomatitis virus populations. J. Virol. *82*, 12589–12590.

Oberste, M.S., Peñaranda, S., and Pallansch, M.A. (2004). RNA recombination plays a major role in genomic change during circulation of coxsackie B viruses. J. Virol. *78*, 2948–2955.

Park, J.M., Muñoz, E., and Deem, M.W. (2010). Quasispecies theory for finite populations. Phys. Rev. E. Stat. Nonlin. Soft Matter Phys. *81*, 011902.

Perales, C., Mateo, R., Mateu, M.G., and Domingo, E. (2007). Insights into RNA virus mutant spectrum and lethal mutagenesis events: replicative interference and complementation by multiple point mutants. J. Mol. Biol. *369*, 985–1000.

Perales, C., Lorenzo-Redondo, R., López-Galíndez, C., Martínez, M.A., and Domingo, E. (2010). Mutant spectra in virus behavior. Future Virology *5*, https://doi.org/10.2217/fvl.10.61

Perales, C., Martín, V., and Domingo, E. (2011). Lethal mutagenesis of viruses. Curr Opin Virol *1*, 419–422.

Pfeiffer, J.K., and Kirkegaard, K. (2003). A single mutation in poliovirus RNA-dependent RNA polymerase confers resistance to mutagenic nucleotide analogs via increased fidelity. Proc. Natl. Acad. Sci. U.S.A. *100*, 7289–7294.

Pfeiffer, J.K., and Kirkegaard, K. (2005). Increased fidelity reduces poliovirus fitness and virulence under selective pressure in mice. PLoS Pathog. *1*, e11.

Pfeiffer, J.K., and Kirkegaard, K. (2006). Bottleneck-mediated quasispecies restriction during spread of an RNA virus from inoculation site to brain. Proc. Natl. Acad. Sci. U.S.A. *103*, 5520–5525.

Prosperi, M.C., and Salemi, M. (2012). QuRe: software for viral quasispecies reconstruction from next-generation sequencing data. Bioinformatics *28*, 132–133.

Quer, J., Esteban, J.I., Cos, J., Sauleda, S., Ocaña, L., Martell, M., Otero, T., Cubero, M., Palou, E., Murillo, P., Esteban, R., and Guàrdia, J. (2005). Effect of bottlenecking on evolution of the nonstructural protein 3 gene of hepatitis C virus during sexually transmitted acute resolving infection. J. Virol. *79*, 15131–15141.

Racaniello, V.R., and Baltimore, D. (1981a). Cloned poliovirus complementary DNA is infectious in mammalian cells. Science *214*, 916–919.

Racaniello, V.R., and Baltimore, D. (1981b). Molecular cloning of poliovirus cDNA and determination of the complete nucleotide sequence of the viral genome. Proc. Natl. Acad. Sci. U.S.A. *78*, 4887–4891.

Rakoto-Andrianarivelo, M., Guillot, S., Iber, J., Balanant, J., Blondel, B., Riquet, F., Martin, J., Kew, O., Randriamanalina, B., Razafinimpiasa, L., Rousset, D., and Delpeyroux, F. (2007). Co-circulation and evolution of polioviruses and species C enteroviruses in a district of Madagascar. PLoS Pathog. *3*, e191.

Rakoto-Andrianarivelo, M., Gumede, N., Jegouic, S., Balanant, J., Andriamamonjy, S.N., Rabemanantsoa, S., Birmingham, M., Randriamanalina, B., Nkolomoni, L., Venter, M., Schoub, B.D., Delpeyroux, F., and Reynes, J.M. (2008). Reemergence of recombinant vaccine-derived poliovirus outbreak in Madagascar. J. Infect. Dis. *197*, 1427–1435.

Riquet, F.B., Blanchard, C., Jegouic, S., Balanant, J., Guillot, S., Vibet, M.A., Rakoto-Andrianarivelo, M., and Delpeyroux, F. (2008). Impact of exogenous sequences on the characteristics of an epidemic type 2 recombinant vaccine-derived poliovirus. J. Virol. *82*, 8927–8932.

Rousset, D., Rakoto-Andrianarivelo, M., Razafindratsimandresy, R., Randriamanalina, B., Guillot, S., Balanant, J., Mauclère, P., and Delpeyroux, F. (2003). Recombinant vaccine-derived poliovirus in Madagascar. Emerg. Infect. Dis. *9*, 885–887.

Saakian, D.B., and Hu, C.K. (2006). Exact solution of the Eigen model with general fitness functions and degradation rates. Proc. Natl. Acad. Sci. U.S.A. *103*, 4935–4939.

Saakian, D.B., Muñoz, E., Hu, C.K., and Deem, M.W. (2006). Quasispecies theory for multiple-peak fitness landscapes. Phys. Rev. E. Stat. Nonlin. Soft Matter Phys. *73*, 041913.

Sadeghipour, S., and McMinn, P.C. (2013). A study of the virulence in mice of high copying fidelity variants of human enterovirus 71. Virus Res. *176*, 265–272.

Sanjuán, R., Moya, A., and Elena, S.F. (2004a). The contribution of epistasis to the architecture of fitness in an RNA virus. Proc. Natl. Acad. Sci. U.S.A. *101*, 15376–15379.

Sanjuán, R., Moya, A., and Elena, S.F. (2004b). The distribution of fitness effects caused by single-nucleotide substitutions in an RNA virus. Proc. Natl. Acad. Sci. *101*, 8396–8401.

Sanjuán, R., Nebot, M.R., Chirico, N., Mansky, L.M., and Belshaw, R. (2010). Viral mutation rates. J. Virol. *84*, 9733–9748.

Sanjuán, R., Lázaro, E., and Vignuzzi, M. (2012). Biomedical implications of viral mutation and evolution. Future Virology *7*, https://doi.org/10.2217/fvl.12.19

Sanjuán, R. (2012). From molecular genetics to phylodynamics: evolutionary relevance of mutation rates across viruses. PLoS Pathog. *8*, e1002685.

Sanz-Ramos, M., Díaz-San Segundo, F., Escarmís, C., Domingo, E., and Sevilla, N. (2008). Hidden virulence determinants in a viral quasispecies in vivo. J. Virol. *82*, 10465–10476.

Sardanyés, J., Elena, S.F., and Solé, R.V. (2008). Simple quasispecies models for the survival-of-the-flattest effect: The role of space. J. Theor. Biol. *250*, 560–568.

Sardanyés, J., Solé, R.V., and Elena, S.F. (2009). Replication mode and landscape topology differentially affect RNA virus mutational load and robustness. J. Virol. *83*, 12579–12589.

Sardanyés, J., and Elena, S.F. (2011). Quasispecies spatial models for RNA viruses with different replication modes and infection strategies. PLOS ONE *6*, e24884.

Sardanyés, J., Simó, C., Martínez, R., Solé, R.V., and Elena, S.F. (2014). Variability in mutational fitness effects prevents full lethal transitions in large quasispecies populations. Sci. Rep. *4*, 4625..

Schibler, M., Gerlach, D., Martinez, Y., Belle, S.V., Turin, L., Kaiser, L., and Tapparel, C. (2012). Experimental human rhinovirus and enterovirus interspecies recombination. J. Gen. Virol. *93*, 93–101.

Sedivy, J.M., Capone, J.P., RajBhandary, U.L., and Sharp, P.A. (1987). An inducible mammalian amber suppressor: propagation of a poliovirus mutant. Cell *50*, 379–389.

Selleri, M., Piralla, A., Rozera, G., Giombini, E., Bartolini, B., Abbate, I., Campanini, G., Rovida, F., Dossena, L., Capobianchi, M.R., and Baldanti, F. (2013). Detection of haemagglutinin D222 polymorphisms in influenza A(H1N1)pdm09-infected patients by ultra-deep pyrosequencing. Clin. Microbiol. Infect. *19*, 668-673.

Semler, B.L., Dorner, A.J., and Wimmer, E. (1984). Production of infectious poliovirus from cloned cDNA is dramatically increased by SV40 transcription and replication signals. Nucleic Acids Res. *12*, 5123–5141.

Simmonds, P., and Welch, J. (2006). Frequency and dynamics of recombination within different species of human enteroviruses. J. Virol. *80*, 483–493.
Simon-Loriere, E., and Holmes, E.C. (2011). Why do RNA viruses recombine? Nat. Rev. Microbiol. *9*, 617–626.
Smith, D.R., Adams, A.P., Kenney, J.L., Wang, E., and Weaver, S.C. (2008). Venezuelan equine encephalitis virus in the mosquito vector Aedes taeniorhynchus: infection initiated by a small number of susceptible epithelial cells and a population bottleneck. Virology *372*, 176–186.
Smith, J.M. (1970). Natural selection and the concept of a protein space. Nature *225*, 563–564.
Smith-Tsurkan, S.D., Wilke, C.O., and Novella, I.S. (2010). Incongruent fitness landscapes, not tradeoffs, dominate the adaptation of vesicular stomatitis virus to novel host types. J. Gen. Virol. *91*, 1484–1493.
Smura, T.P., Junttila, N., Blomqvist, S., Norder, H., Kaijalainen, S., Paananen, A., Magnius, L.O., Hovi, T., and Roivainen, M. (2007). Enterovirus 94, a proposed new serotype in human enterovirus species D. J. Gen. Virol. *88*, 849–858.
Teng, M.N., Oldstone, M.B., and de la Torre, J.C. (1996). Suppression of lymphocytic choriomeningitis virus-induced growth hormone deficiency syndrome by disease-negative virus variants. Virology *223*, 113-119.
Thompson, A.A., Albertini, R.A., and Peersen, O.B. (2007). Stabilization of poliovirus polymerase by NTP binding and fingers-thumb interactions. J. Mol. Biol. *366*, 1459–1474.
Tokarz, R., Williams, S.H., Sameroff, S., Sanchez Leon, M., Jain, K., and Lipkin, W.I. (2014). Virome analysis of Amblyomma americanum, Dermacentor variabilis, and Ixodes scapularis ticks reveals novel highly divergent vertebrate and invertebrate viruses. J. Virol. *88*, 11480–11492.
Töpfer, A., Höper, D., Blome, S., Beer, M., Beerenwinkel, N., Ruggli, N., and Leifer, I. (2013). Sequencing approach to analyze the role of quasispecies for classical swine fever. Virology *438*, 14–19.
van der Sanden, S., van Eek, J., Martin, D.P., van der Avoort, H., Vennema, H., and Koopmans, M. (2011). Detection of recombination breakpoints in the genomes of human enterovirus 71 strains isolated in the Netherlands in epidemic and non-epidemic years, 1963-2010. Infect. Genet. Evol. *11*, 886–894.
van der Werf, S., Bradley, J., Wimmer, E., Studier, F.W., and Dunn, J.J. (1986). Synthesis of infectious poliovirus RNA by purified T7 RNA polymerase. Proc. Natl. Acad. Sci. U.S.A. *83*, 2330–2334.
Vasilakis, N., Forrester, N.L., Palacios, G., Nasar, F., Savji, N., Rossi, S.L., Guzman, H., Wood, T.G., Popov, V., Gorchakov, R., González, A.V., Haddow, A.D., Watts, D.M., da Rosa, A.P., Weaver, S.C., Lipkin, W.I., and Tesh, R.B. (2013). Negevirus: a proposed new taxon of insect-specific viruses with wide geographic distribution. J. Virol. *87*, 2475–2488.
Victoria, J.G., Kapoor, A., Li, L., Blinkova, O., Slikas, B., Wang, C., Naeem, A., Zaidi, S., and Delwart, E. (2009). Metagenomic analyses of viruses in stool samples from children with acute flaccid paralysis. J. Virol. *83*, 4642–4651.
Vignuzzi, M., Stone, J.K., Arnold, J.J., Cameron, C.E., and Andino, R. (2006). Quasispecies diversity determines pathogenesis through cooperative interactions in a viral population. Nature *439*, 344–348.
Vignuzzi, M., Wendt, E., and Andino, R. (2008). Engineering attenuated virus vaccines by controlling replication fidelity. Nat. Med. *14*, 154–161.
Wilke, C.O., Wang, J.L., Ofria, C., Lenski, R.E., and Adami, C. (2001). Evolution of digital organisms at high mutation rates leads to survival of the flattest. Nature *412*, 331–333.
Wilke, C.O., and Ronnewinkel, C. (2001). Dynamic fitness landscapes: expansions for small mutation rates. Physica A: Statistical Mechanics and its Applications *290*, 475-490.
Wilke, C.O., Ronnewinkel, C., and Martinetz, T. (2001a). Dynamic fitness landscapes in molecular evolution. Physics Reports *349*,395-446.
Wilke, C.O. (2001b). Adaptive evolution on neutral networks. Bull. Math. Biol. *63*, 715–730.
Wilke, C.O., Reissig, D.D., and Novella, I.S. (2004). Replication at periodically changing multiplicity of infection promotes stable coexistence of competing viral populations. Evolution *58*, 900–905.
Wilke, C.O. (2005). Quasispecies theory in the context of population genetics. BMC Evol. Biol. *5*, 44.
Wilke, C.O., Forster, R., and Novella, I.S. (2006). Quasispecies in time-dependent environments. Curr. Top. Microbiol. Immunol. *299*, 33–50.
Wilm, A., Aw, P.P., Bertrand, D., Yeo, G.H., Ong, S.H., Wong, C.H., Khor, C.C., Petric, R., Hibberd, M.L., and Nagarajan, N. (2012). LoFreq: a sequence-quality aware, ultra-sensitive variant caller for uncovering cell-population heterogeneity from high-throughput sequencing datasets. Nucleic Acids Res. *40*, 11189-11201.
Wimmer, E., and Paul, A.V. (2011). Synthetic poliovirus and other designer viruses: what have we learned from them? Annu. Rev. Microbiol. *65*, 583–609.

Worobey, M., and Holmes, E.C. (1999). Evolutionary aspects of recombination in RNA viruses. J. Gen. Virol. *80* 2535–2543.
Wright, C.F., Morelli, M.J., Thébaud, G., Knowles, N.J., Herzyk, P., Paton, D.J., Haydon, D.T., and King, D.P. (2011). Beyond the consensus: dissecting within-host viral population diversity of foot-and-mouth disease virus by using next-generation genome sequencing. J. Virol. *85,* 2266-2275.
Xiao, Y., Rouzine, I.M., Bianco, S., Acevedo, A., Goldstein, E.F., Farkov, M., Brodsky, L., and Andino, R. (2016). RNA Recombination Enhances Adaptability and Is Required for Virus Spread and Virulence. Cell Host Microbe *19,* 493–503.
Yang, X., Welch, J.L., Arnold, J.J., and Boehr, D.D. (2010). Long-range interaction networks in the function and fidelity of poliovirus RNA-dependent RNA polymerase studied by nuclear magnetic resonance. Biochemistry *49,* 9361–9371.
Yang, X., Smidansky, E.D., Maksimchuk, K.R., Lum, D., Welch, J.L., Arnold, J.J., Cameron, C.E., and Boehr, D.D. (2012). Motif D of viral RNA-dependent RNA polymerases determines efficiency and fidelity of nucleotide addition. Structure *20,* 1519–1527.
Yang, C., Skiena, S., Futcher, B., Mueller, S., and Wimmer, E. (2013a). Deliberate reduction of hemagglutinin and neuraminidase expression of influenza virus leads to an ultraprotective live vaccine in mice. Proc. Natl. Acad. Sci. U.S.A. *110,* 9481–9486.
Yang, X., Charlebois, P., Macalalad, A., Henn, M.R., and Zody, M.C. (2013b). V-Phaser 2: variant inference for viral populations. BMC Genomics *14,* 674.
Yozwiak, N.L., Skewes-Cox, P., Gordon, A., Saborio, S., Kuan, G., Balmaseda, A., Ganem, D., Harris, E., and DeRisi, J.L. (2010). Human enterovirus 109: a novel interspecies recombinant enterovirus isolated from a case of acute pediatric respiratory illness in Nicaragua. J. Virol. *84,* 9047–9058.
Yuste, E., Sánchez-Palomino, S., Casado, C., Domingo, E., and López-Galíndez, C. (1999). Drastic fitness loss in human immunodeficiency virus type 1 upon serial bottleneck events. J. Virol. *73,* 2745–2751.
Yuste, E., Bordería, A.V., Domingo, E., and López-Galíndez, C. (2005). Few mutations in the 5′ leader region mediate fitness recovery of debilitated human immunodeficiency type 1 viruses. J. Virol. *79,* 5421–5427.
Zagordi, O., Geyrhofer, L., Roth, V., and Beerenwinkel, N. (2010). Deep sequencing of a genetically heterogeneous sample: local haplotype reconstruction and read error correction. J. Comput. Biol. *17,* 417–428.
Zagordi, O., Bhattacharya, A., Eriksson, N., and Beerenwinkel, N. (2011). ShoRAH: estimating the genetic diversity of a mixed sample from next-generation sequencing data. BMC Bioinformatics *12,* 119.
Zeng, J., Wang, H., Xie, X., Yang, D., Zhou, G., and Yu, L. (2013). An increased replication fidelity mutant of foot-and-mouth disease virus retains fitness in vitro and virulence in vivo. Antiviral Res. *100,* 1–7.
Zeng, J., Wang, H., Xie, X., Li, C., Zhou, G., Yang, D., and Yu, L. (2014). Ribavirin-resistant variants of foot-and-mouth disease virus: the effect of restricted quasispecies diversity on viral virulence. J. Virol. *88,* 4008-4020.

Enterovirus Control of Cytoplasmic RNA Granules

6

Richard E. Lloyd

Department of Molecular Virology and Microbiology, Baylor College of Medicine, Houston, TX, USA.

Correspondence: rlloyd@bcm.edu

https://doi.org/10.21775/9781910190739.06

Abstract

RNA granules are dynamic structures in cells that are closely linked to repression of gene expression and comprise essential parts of the life cycle of mRNPs. The two general types of RNA granules in somatic cells are stress granules (SGs) and processing bodies (P-bodies, PBs), both of which are manipulated by many types of viruses. Enteroviruses such as poliovirus (PV) and coxsackievirus B3 (CVB3) induce stress granule formation but almost immediately block their formation during the course of infection. PBs are constitutively present, and *Enterovirus* infection results in their complete dispersal. This review discusses these processes and the current understanding of the underlying mechanisms. In addition, the review discusses reasons viruses control RNA granules, including accumulating data suggesting linkage between RNA granule formation and innate immune sensing and activation.

Cytoplasmic RNA granules

mRNAs are highly regulated post-transcriptionally by changing their mRNP protein compositions in order to control splicing, export, translation regulation, some cellular localization, and turnover. mRNA translation in particular is highly regulated and a large proportion of mRNAs are undergoing translational restriction or outright silencing at any time in cells due to fluctuating mRNP compositions. These pools of silenced transcripts, often resulting from microRNA repression, are constantly shifting as cells adapt to changing conditions and signals. Cytoplasmic RNA granules are compartments where translationally silenced transcripts are temporarily stored and are thus seen as extensions of translational silencing mechanisms. The two major classes of cytoplasmic RNA granules in somatic cells are known as stress granules (SGs) and processing bodies (P-bodies, PBs). Several types of nuclear RNA granules are known (e.g. cajal bodies, nuclear speckles, paraspeckles), but are distinct and will not be discussed here.

Stress granules are normally not observed in rapidly growing cells but quickly organize when various types of environmental stress is applied to cells, including oxidative stress, heat shock, nutrient deprivation, etc. all of which result in decreased global translation rates. Stress granules are identified as condensates that contain high concentrations of stalled translation initiation complexes. Thus, the most diagnostic markers for stress granules are translation initiation factors and components involved in mRNA-ribosome binding and scanning. These include eIF4E, eIF4G, eIF4A, eIF3, PABP, the small 40S ribosome and mRNA itself. SGs also contain a plethora of mRNA-binding proteins (RBPs), likely any factor that associates with mRNPs may be incorporated into SGs along with their associated mRNAs, however many of these factors will not have important roles in SG biology. There is a subset of other factors known to nucleate SGs that play various key roles in SG assembly and disassembly. These proteins are RNA binding factors that can normally reside in the cytoplasm or nucleus, and may play multiple roles in gene expression regulation. The list of key SG-nucleating factors includes T-cell-restricted intracellular antigen 1(Tia1),

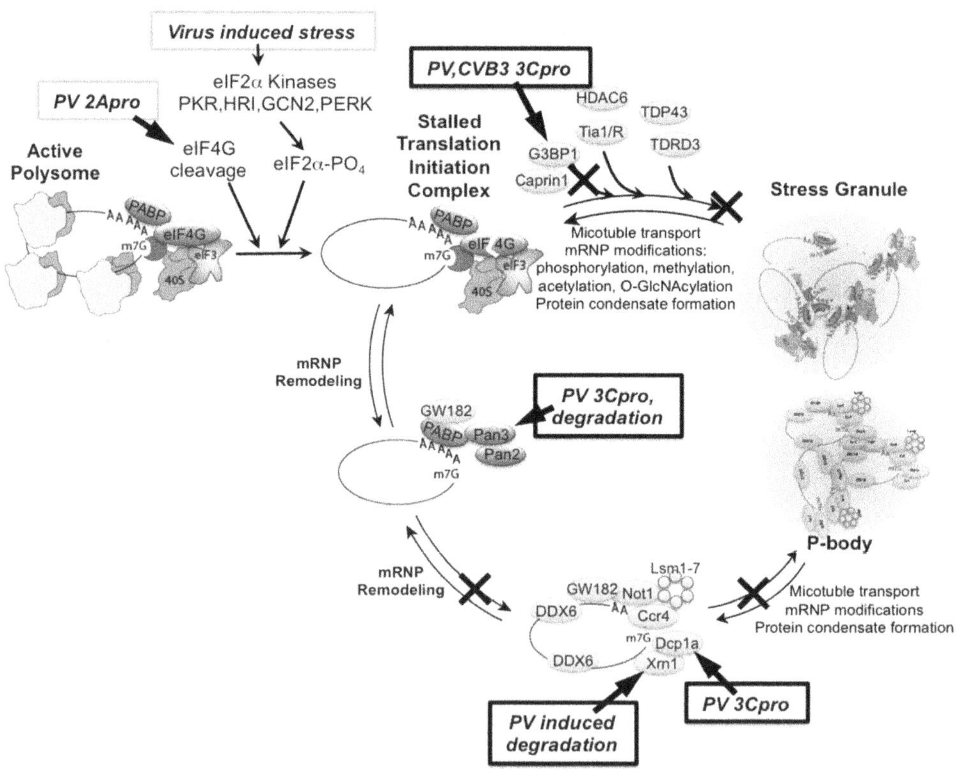

Figure 6.1 Stress granule assembly and factors targeted by Enteroviruses. Active polysomes are disassembled by cleavage of eIF4G and PABP as well as rising concentrations of phosphorylated eIF2α from activation of PKR and likely other eIF2α kinases. The accumulating stalled translation initiation complexes are then incorporated into stress granules or also possibly undergo alternate mRNP remodelling and are incorporated into P-bodies (see Fig. 6.3). SG assembly additionally requires condensation of SG-nucleating proteins such as G3BP1, Tia1, TiaR, TDP43, TDRD3, plus microtubule transport and post-translational modifications on many mRNP proteins. All steps are reversible as indicated. PV blocks both SG assembly and disperses PBs and directly or indirectly degrades the indicated factors.

RasGAP-SH3 domain binding protein 1 and 2 (G3BP1 and G3BP2), Tristetraprolin (TTP), caprin1, and TDP-43, among others (Fig. 6.1). While many of these proteins are critical for SG function, their usage as markers to identify SGs is arguably overused and often can be misleading in virus systems since some viruses co-opt some of these factors into novel types of foci.

Processing bodies are different from SGs in that some number of PBs are usually constitutively present in somatic cells under normal growth conditions. Similar to SGs, they do respond to stress signals by increasing size and number, though often not as dramatically. PBs have a distinct mRNP constituency from SGs, in that the translationally silenced mRNPs have been stripped of ribosome subunits and most translation initiation factors before organization into PBs. Some RBPs and associated proteins can be found in both SGs and PBs. Importantly, PBs are enriched for mRNA decay machinery such as Xrn1, decapping factors Dcp1a, Dcp2, EDC3,4 and poly(A) nucleases PAN2, PAN3, Ccr4 and Not1 (Fig. 6.2).

It has been proposed that a cytoplasmic mRNA cycle exists where active and silenced mRNPs are in equilibrium and mRNPs rapidly shuttle between active polysomes and the silenced compartments comprised of SGs and PBs. As a part of this cycle, mRNPs are proposed to shuttle between SGs and PBs, which can often be observed docked together in cells. In this regard, several components that are shared between SGs and PBs, such as TTP, Ago2, APOBEC3, eIF4E and others may play important mRNP shuttling roles (Kedersha and Anderson, 2007; Kedersha *et al.*, 2005). This is an important but understudied area.

Mechanisms of stress granule assembly

Stress granules contain stalled translation initiation complexes, so the canonical pathway for SG assembly starts with triggering translation inhibition via several pathways. The most commonly described is through activation of one or more of the stress activated eIF2α-kinases, PKR, HRI, GCN2 or PERK (haem-regulated kinase, HRI; general control non- depressible 2 kinase, GCN2; double-stranded RNA (dsRNA)-activated protein kinase R, PKR; and PKR-like endoplasmic reticulum kinase, PERK). All four of these factors sense various types of stress signals, resulting in activation of their kinase functions and result in phosphorylation of eIF2α on serine-51. This phosphorylation results in global translation repression through depletion of eIF2•GTP•met-$tRNA_i^{met}$ that is required for translation initiation on all transcripts that initiate with AUG codons. The result is accumulation of stalled 43S and 48S translation initiation complexes that ultimately must be stored in the SG compartment (Fig. 6.1). In this way SGs can be seen as a temporary 'storage closet' for unused or silenced mRNAs. Alternate modes of SG formation can proceed in the absence of eIF2α phosphorylation and may involve inhibition of eIF4A helicase function, cleavage of eIF4GI or other mechanisms (Dang *et al.*, 2006; Emara *et al.*, 2012; Mazroui *et al.*, 2006; Reineke *et al.*, 2012).

The actual aggregation or phase-condensation stage of RNA granule formation is poorly understood but requires a series of events. These include transport of mRNPs on microtubule networks with the aid of dynein and kinesin motor proteins, protein-protein phase condensation functions of key nucleating proteins, as well as post translational modification of multiple mRNP proteins via phosphorylation, acetylation, methylation and O-linked N-acetylglucosamine (Chernov *et al.*, 2009; Kedersha and Anderson, 2007; Kedersha *et al.*,

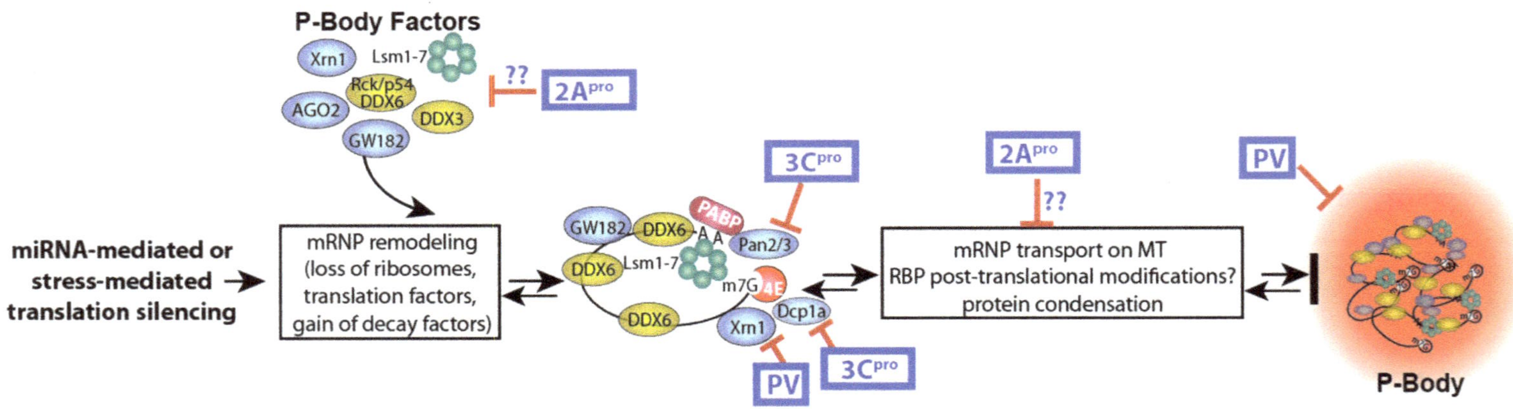

Figure 6.2 P-body assembly and factors targeted by Enteroviruses. Poliovirus strongly counteracts PB assembly. Mechanisms that govern PB assembly are poorly understood. PV $3C^{pro}$ cleaves Dcp1a and induces degradation of Pan3. Xrn1 is also targeted by rapid proteasomal degradation. PV $2A^{pro}$ also represses PB, though no $2A^{pro}$ molecular targets that affect PB function are known. $2A^{pro}$ could cleave additional P-body factors or factors required for microtubule transport or post-translational modifications that promote protein condensation. Only a small subset of known constituents of SGs and PBs are depicted.

2005; Kwon *et al.*, 2007; Loschi *et al.*, 2009; Nadezhdina *et al.*, 2010; Ohn *et al.*, 2008; Tsai *et al.*, 2009; Xie and Denman, 2011). The post-translational protein modifications are proposed to be key in changing protein–protein interactions that drive cytoplasmic condensate formation. For SGs, this occurs on a set of key proteins that play nucleating roles in SG formation, owing to the presence of aggregation or prion-like low amino acid complexity motifs in their sequences (Kedersha *et al.*, 2013). Key among SG nucleating proteins are Tia1, G3BP1 and G3BP2, TTP, caprin1, and TDP-43 (Aulas *et al.*, 2012; Gilks *et al.*, 2004; Tourrière *et al.*, 2003). Expression of truncated forms of Tia1 or especially G3BP1 results in dominant negative inhibition of SG formation (Gilks *et al.*, 2004; Tourrière *et al.*, 2003). G3BP1 is phosphorylated by casein kinase 2 to repress SG formation (Reineke *et al.*, 2017) and dephosphorylated by unknown phosphatase to promote SG formation. G3BP is also methylated by protein arginine methyltransferase 1 and protein arginine methyltransferase 5, also to repress SG formation. In fact, G3BP1 methylation is rapidly decreased during cell stress to facilitate SG formation (Tsai *et al.*, 2016).

It is important to note that not all SGs are compositionally equivalent, and SGs constituents vary depending on the mode of stress that induced them. The variants of SGs all share markers that define SG function as repositories of stalled translation complexes, e.g. translation initiation factors, 40S ribosome subunits, mRNA. But other factors may be unique, for example, heat shock induced stress granules contain heat shock protein 27 (hsp27), but hsp27 is absent in arsenite (Ars)-induced SGs (Kedersha *et al.*, 1999; Gilks *et al.*, 2004; Piotrowska *et al.*, 2010). Virus infection produces unique types of cell stress and can induce SG to form. Some poliovirus-induced SGs contain Sam68 which is not found in heat shock-induced SGs (Piotrowska *et al.*, 2010).

Enterovirus relationships with stress granules: antagonism rules

Viruses manipulate host gene expression to create conditions conducive for virus replication. As a result, virus infection produces new stresses on cells at multiple levels and frequently exerts mechanisms of translation control. These insults are expected to trigger RNA granule flux and formation and many types of viruses are known to modulate SGs [reviewed in (Reineke and Lloyd, 2013; Tsai and Lloyd, 2014)]. In most cases described so far, viruses exert stringent control over the dynamic fluxes that form SGs and many viruses also regulate PBs. Some plus-strand viruses such as alphaviruses, flaviviruses and hepatitis C virus co-opt key SG nucleating proteins such as Tia1 or G3BP1 into new roles supporting virus replication [reviewed in Kedersha and Anderson (2007); Kedersha *et al.* (2005); Lloyd (2012); Reineke and Lloyd (2013)]. The Enteroviruses poliovirus (PV) and coxsackievirus B3 (CVB3) are different in that SG proteins are not co-opted, but instead are destroyed through cleavage. Enteroviruses are excellent probes to study RNA granule function in that they strongly regulate both SGs and PBs.

Overview of *Enterovirus* infection: stress granule induction and then destruction

Enteroviruses such as poliovirus and CVB3 exert powerful controls over a large range of host processes and since RNA granule formation is multistage and multifactorial, these viruses may be expected to control SGs through several channels. Since SG are principally repositories of stalled translation complexes, virus-induced mechanisms that regulate host translation are key in their formation. PV and CVB3 shut down host translation aggressively via cleavage of eIF4GI and eIF4GII using $2A^{pro}$ and through induction of cellular proteases, both which cleave this key initiation factor (Dang *et al.*, 2006; Emara *et al.*, 2012; Etchison *et al.*, 1982; Gradi *et al.*, 1998; Mazroui *et al.*, 2006; Reineke *et al.*, 2012; Zamora *et al.*, 2002). These events result in accumulation of inactive translation machinery that is later co-opted by viral mRNA, but will also trigger SG formation. Thus, it was first reported that poliovirus infection induced stress granule formation by 3 hours post infection (hpi), which was linked to cleavage of eIF4G (Mazroui *et al.*, 2006) (Fig. 6.1). Curiously, the authors did not follow the infection further into the main biosynthetic replicative phase, thus the remainder of an interesting story remained undiscovered initially.

In fact, poliovirus infection first causes SG formation early in infection and this phase is rapidly replaced by strident blockage of SG formation by the midphase of the replicative cycle. Infection of different cell types indicated the percentage of infected cells that achieved SG formation early in infection was variable, ranging from 15–80% depending on the cell line infection (White *et al.*, 2007). This may be reflected in the permissiveness of the infection, as infections that display somewhat slower kinetics often (but not always) produce higher levels of SGs. Any SGs that form by 3 hpi then rapidly begin to dissipate and are generally absent by 4–5 hpi in the infected cell lines examined. The loss of virus-induced SGs in cells correlated with appearance of an activity that blocks SG formation when exogenous stressors such as sodium arsenite are applied. Thus, the mechanism(s) that drives SG formation is strongly countered by one or more virus functions that result in disassembly of SGs and prevents their formation (White *et al.*, 2007). These salient features of SG formation and repression, as expected, were also later shown to occur with CVB3 and EV71 during infection (Fung *et al.*, 2013; Wu *et al.*, 2014b).

Viral proteases induce stress granules and disperse stress granules

As noted above, eIF4G cleavage by viral 2A proteinase ($2A^{pro}$) is strongly linked to SG formation early in PV infection. $2A^{pro}$ and 3C proteinase ($3C^{pro}$)-mediated cleavage of PABP contributes strongly to host translation shutoff as well (Kuyumcu-Martinez *et al.*, 2004; Lloyd, 2012; Reineke and Lloyd, 2013) but the relative roles of cleavage of eIF4GI versus PABP in inducing SGs have not been determined. As expected, the SG-inducing role of $2A^{pro}$ has been confirmed by expression of individual viral PV or CVB3 proteins in cells (Dougherty *et al.*, 2015; Wu *et al.*, 2014b). Further, since PKR (and other eIF2α kinases) is a canonical activator of stress granule assembly, it likely plays an ancillary role in driving SG assembly early in infection. A steady increase in PKR activation and eIF2α phosphorylation are well documented in Enterovirus infection, even though the virus utilizes mechanisms, to preclude eIF2α-phosphorylation from getting out of check and controlling viral translation until late in the replication cycle (Black *et al.*, 1993; O'Neill and Racaniello, 1989; White *et al.*, 2011) (Fig. 6.1).

How is SG assembly blocked by Enteroviruses? Screening the fate of key SG nucleating proteins during infection revealed that G3BP1, but not Tia1 or TIAR, was cleaved during poliovirus infection. This cleavage is catalysed by PV $3C^{pro}$, which efficiently cleaves G3BP1 *in vitro* (Fig. 6.1). The cleavage site results in separation of the N-terminal NTF2-acidic-PxxP protein interaction motifs of G3BP from the adjacent RNA binding motifs (RRM and RGG) (White *et al.*, 2007) (Fig. 6.3). As expected, very similar results were recently

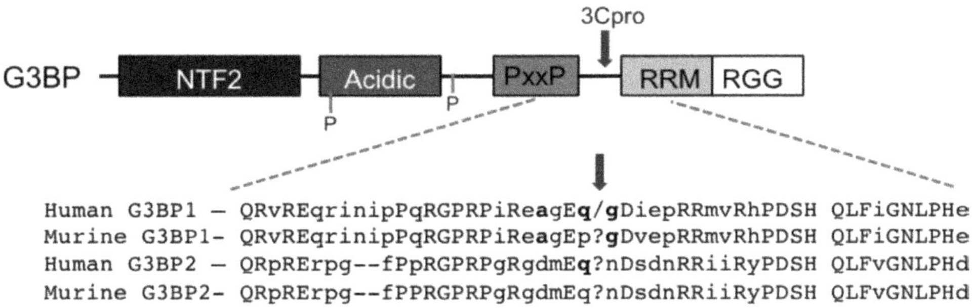

Figure 6.3 3C^pro^ cleavage site region in human G3BP1 is absent in G3BP2 and murine G3BP1 and G3BP2. Schematic shows the conserved domains of G3BP and location of 3C^{pro} cleavage site. That region is expanded and sequence alignment shown comparing human and mouse G3BP1 and G3BP2 amino acid sequences. 3C^{pro} cleavage requires alanine in P4 position, glutamine at P1 and glycine at P1' for efficient cleavage. Those residues are highlighted in boldface. Only human G3BP1 has all required residues for 3C^{pro} cleavage.

reported for CVB3 infection and EV71, with SG formation early in infection followed by disassembly in conjunction with G3BP1 cleavage at the same site (Fung *et al.*, 2013; Wu *et al.*, 2014b). Further, a more distantly related picornavirus, encephalomyocarditis virus (EMCV), also disrupts SGs by cleaving human G3BP1 at the same site (Ng *et al.*, 2013), thus two distinct genera of picornaviruses can target G3BP1 to disrupt SG.

How does G3BP cleavage block stress granule assembly?

The cleavage of G3BP1 observed during PV infection focused new attention on the role of G3BP in SG assembly, and has resulted in new insights. Recent reports indicate that both G3BP1 and its homologue G3BP2 are important in SG formation and may provide some duplicity of function. Both G3BP1 and G3BP2 share the same five protein domains and the N-terminal NTF2 domain in particular is highly conserved (Fig. 6.3). G3BP1 and G3BP2 may dimerize through interactions of the NTF2 domain, and because of the high homology, can form homodimers and heterodimers (Matsuki *et al.*, 2013). Either protein can induce SGs when overexpressed and siRNA knockdown of both is required to most effectively repress SG formation (Kedersha *et al.*, 2005; Matsuki *et al.*, 2013; Reineke *et al.*, 2012; Tourrière *et al.*, 2003). siRNA depletion of G3BP1 resulted only partial reduction of granules containing stalled initiation complexes after arsenite treatment (Kedersha *et al.*, 2013; White *et al.*, 2007) and numerous other approaches have indicated at least a partial duplicity in function between G3BP1 and G3BP2 (Reineke *et al.*, 2015). However, work with PV indicates that the role of G3BP1 is dominant in SG function. The cleavage site on G3BP1 was mapped to the glutamine–glycine bond at Q326. This site is not conserved in G3BP2 (Fig. 6.3) and G3BP2 is not cleaved during PV infection or by incubation with 3C^{pro} (Dougherty *et al.*, 2015). Mutation of the G3BP1 cleavage site to Q326E renders G3BP non-cleavable by 3C^{pro}. Experiments with introduction of cleavage resistant G3BP^{Q326E} back into cells resulted in restoration of the ability to maintain SGs during infection (White *et al.*, 2007). Further, while there is duplicity in G3BP function, differences between G3BP1 and G3BP2 are emerging, as only the latter interacts with and is phosphorylated by protein kinase Cα (Kobayashi *et al.*, 2012). Kinases that modify G3BP1 have not yet been reported but include casein kinase II (CKII) since the serine149 phosphorylation site that is associated with prevention of SG assembly by G3BP1 lies within a canonical CKII recognition site (Tourrière *et al.*, 2003; Reineke *et al.*, 2017).

Virus cleavage of G3BP1 is closely linked to SG destruction, but what is the real underlying mechanism? Such questions require understanding of the mechanism by which G3BP1 drives SG assembly, which is poorly understood. G3BP1 is thought to function as a physical aggregator that links mRNPs together via its protein–protein interaction dimerization domains and its PxxP domain that has a low complexity amino acid diversity (LC) and is intrinsically disordered (ID). G3BP1 binds mRNA via the C-terminal RRM/RGG domains (Fig. 6.3), but the breadth or extent of cellular mRNAs it interacts with is unknown. Similar aggregation functions are proposed for LC/ID prion-like regions of Tia1 and TiaR. These proteins presumably bind different subsets of cell mRNAs than G3BP, however this hypothesis has not been tested. Indeed, one can assemble a list of over 20 factors that have LC/ID regions and play roles in SG nucleation (Kedersha *et al.*, 2013).

So how is it that cleavage of only one of the multiple known SG aggregation proteins has such impact on SG formation? Is the production of G3BP1 cleavage products more important than depletion of functional G3BP1 in the cell? Several reports indicate that G3BP cleavage products are more effective in blocking SGs than siRNA knockdown of G3BP1. Early characterization of G3BP1 via expression of truncated regions of the protein indicated that certain truncations of G3BP1 are dominant negative inhibitors of SG formation (Tourrière *et al.*, 2003). Some of these inhibitory truncations were quite different than fragments produced during virus infection. More recent work indicates that the cleavage fragments of G3BP1 generated by CVB3 $3C^{pro}$, particularly the C-terminal RNA binding domain (identical to the PV-generated cleavage fragments) can dominantly repress SG formation (Fung *et al.*, 2013). Our work indicates that removal of the N-terminal NTF2-like domain is sufficient to render G3BP1 incapable of nucleating SGs and produce a dominant negative (Reineke and Lloyd, 2015). Taken together, various truncations of G3BP1 efficiently repress SG formation. This can be interpreted to suggest that multiple interactions along the length of G3BP are needed to support SG formation. Of course, much more work is required to understand exactly how G3BP1 functions as such a powerful nucleator of SGs, what cofactors it requires and what subset of cellular transcripts it assembles into granules.

Other viral targets that control stress granules?

There is also a possibility that other cell proteins targeted by Enteroviruses may play roles in SG formation. Cleavage of eIF4G helps drive SG formation, but was proposed to play a role in SG disassembly also (Piotrowska *et al.*, 2010). However eIF4G cleavage was excluded as playing a significant role in SG disassembly since both N-terminal and C-terminal $2A^{pro}$ cleavage fragments of eIF4GI can assemble in arsenite-induced SGs under infection conditions where G3BP is not cleaved (White and Lloyd, 2011; White *et al.*, 2007). A likely possibility is that $2A^{pro}$, which drives SG formation by inducing eIF4G cleavage may also play a role in blocking SG formation. Expression of $2A^{pro}$–cherry fusion constructs in HeLa and A549 cells is able to partly inhibit the formation of SGs induced by arsenite stress but much less avidly than $3C^{pro}$ (Dougherty *et al.*, 2015). However, a cellular substrate of $2A^{pro}$ that is targeted in this function has not been identified. Likewise, $3C^{pro}$ may also target other factors that support SG formation to lesser degrees than G3BP1 or via indirect means.

Recently, the autophagy-lysosome pathway and valosin-containing protein (VCP), both key players in the protein quality control (PQC) pathways, were shown to regulate SG degradation in yeast. This indicates that protein quality control systems may survey and/or assist SG dynamics. In mammalian cells with impaired autophagy or VCP function SGs were defective, containing 60S ribosomes and DRIPs that should be normally excluded. This suggests that deregulated autophagy, lysosomal or VCP activities, which occur in several neurodegenerative (VCP-associated) diseases, may alter SG morphology and composition (Buchan *et al.*, 2013; Seguin *et al.*, 2014). Enteroviruses are well known to disrupt and modulate autophagic processes, which is detailed elsewhere in this volume. It is possible that virus disruption of autophagy also plays a role in SG dynamics in infected cells.

Enteroviruses unlink TIA1 aggregation from stress granule formation

Though G3BP1 cleavage is crucial to block SG formation, a seemingly contradictory report indicated Tia1-containing SGs persisted even during late times during PV infection (Piotrowska *et al.*, 2010). Those Tia1-containing granules however were later shown to be smaller than the initial bona fide SGs that form during infection. The persisting Tia1 granules were devoid of multiple components of stalled translation complexes, including eIF3, eIF4A, eIF4E, 40S subunits and most mRNA detectable by poly(A)-directed in situ hybridization (White and Lloyd, 2011). The loss of translation components from Tia1-containing granules correlated with the time post infection that G3BP was cleaved. Thus, aggregation of intact G3BP1 was linked with the functional role of SGs in condensation of translation complexes and not Tia1 aggregation. Therefore poliovirus infection unlinks stress-induced Tia1 aggregation from formation of functional SGs (White and Lloyd, 2011). This result is important because numerous investigators have relied heavily on scoring SG formation by following foci containing one or two so-called 'marker proteins' instead of following the functional translation components that define stalled initiation complexes. Many reports indicate that a common theme during virus infection involves viruses co-opting key SG nucleating proteins into novel foci that are not SGs. Most SG-inducing proteins are RNA-binding proteins with inherent aggregating or phase condensation properties. These can be diverted to novel roles, at the same time preventing SG formation and providing new tools for various steps in virus replication or assembly (Reineke and Lloyd, 2013; Tsai and Lloyd, 2014). These findings also reinforce the notion that mRNP granules differ widely in composition and function and should not be stereotyped based on a few limited factors they have in common.

Other RNA-binding proteins targeted by Enteroviruses

Recent reports have indicated that adenosine-uridine (AU)-rich element RNA binding factor 1 (AUF1), a well described mRNA destabilizing factor, also destabilizes CVB3 and PV viral RNA (Cathcart *et al.*, 2013; Wong *et al.*, 2013). This factor is nuclear-cytoplasmic shuttling, though is retains nuclear distribution until midphase of viral infection, when its distribution becomes partly cytoplasmic and it can inhibit viral translation via interaction with viral IRES (Cathcart *et al.*, 2013). One report indicated that AUF1 is weakly recruited to SGs in CVB3-infected cells (Wu *et al.*, 2014a); however, this may result solely from incorporation of cellular mRNPs containing AUF1 as a constituent. The viral translation-restricting function of AUF1 likely has little impact on viral infection when AUF1 enters into SGs, since viral mRNA does not enter SGs (Piotrowska *et al.*, 2010). Further, PV $3C^{pro}$ and 3CD cleave AUF1 to counteract is restriction activity (Rozovics *et al.*, 2012).

Stress granule inhibition by other viruses in the order Picornavirales

It is expected that all human Enteroviruses will employ identical or similar mechanisms to control RNA granules, since all conserve both 2A and 3Cproteases. But what can be learned from other picornaviruses that lack a 2A protease but retain $3C^{pro}$? Theiler's murine encephalomyelitis virus (TMEV) also blocks SG formation through a mechanism that does not involve G3BP1 cleavage (Borghese and Michiels, 2011). Mouse G3BP1 does not conserve the same Q/G cleavage site found in human G3BP1 so the virus may have evolved alternate mechanisms to evade SGs. However, infections performed in HeLa cells with cleavage-susceptible G3BP1 also failed to cleave G3BP1, so this $3C^{pro}$ has evolved cleavage site specificities that preclude cleavage of both human and mouse G3BP1. Interestingly, the TMEV leader protein, which is not a proteinase, was determined to block SG formation. Ectopic expression of the leader protein could block cells from mounting stress granule responses to arsenite and thapsigargin (Borghese and Michiels, 2011). Similar abilities to block SG formation were observed with recombinant TMEV expressing the L-proteins of Saffold virus and Mengo virus. Partly in contrast with these findings, another *Cardiovirus*, EMCV, was found to disrupt SGs through cleavage of G3BP1 in HeLa cells at the same Q/G bond cleaved by PV (Ng *et al.*, 2013). Cleavage of G3BP1 in murine cells was not reported, thus the role of this additional poliovirus-like mechanism in natural mouse infections is speculative because of the lack of conservation of the cleavage site in murine G3BP1 or G3BP2 (Fig. 6.3). Mengo virus, considered a strain of EMCV, controls SG formation in mouse cells via its leader protein, which can be inactivated in this function by mutation of a zinc finger motif (Langereis *et al.*, 2013) or the Theilo domain (Borghese and Michiels, 2011).

Cricket paralysis virus, a member of the Dicistrovirus family, can also block stress granule formation at early times post infection (2 hpi). Similar to poliovirus, activities that inhibit stress granules formation induced by exogenous stressors also appears later (4 hpi), indicating that stress granule inhibition correlates with and is dependent on increased viral protein expression. Cleavage of the G3BP paralogue Rasputin does not occur during infection, in contrast to poliovirus infection, and the paralogue of Tia1 also remains intact (Khong and Jan, 2011). Viral protein 1A modulates SG assembly and host transcription, through unknown mechanisms that facilitate virus replication (Khong and Jan, 2016).

Enterovirus relationships with P-bodies: rapid destruction

PBs are sites of enriched mRNA decay machinery in the cytoplasm, containing most if not all the major components involved in both 5′ and 3′ directed mRNA decay. These include the poly(A)-nucleases, the exosome that catalyses 3′ mediated decay, plus decapping machinery/Xrn1 complexes that catalyse 5′-end mediated decay (Jonas and Izaurralde, 2013; Teixeira *et al.*, 2005). Since PV mRNA contains no cap and is not protected by VPg (Langereis *et al.*, 2014; Virgen-Slane *et al.*, 2012), it is uniquely susceptible to 5′ end mediated decay that could destroy viral RNA if left unchecked. Thus it is of no surprise that

Enteroviruses also antagonize RNA decay factors, many of which are PB components. Both PV and CVB3 infections cause dispersal and destruction of PBs during infection, which is complete by 4 hpi, roughly at the time that virus biosynthetic activity is approaching its peak levels (Dougherty *et al.*, 2011).

P-body constituent proteins targeted by Enteroviruses

Less is known about the mechanisms of PB assembly than SG assembly; however, decapping complex proteins with low amino acid complexity disordered regions are proposed to play similar roles in complex formation and higher order RNA granule formation (Jonas and Izaurralde, 2013; Kedersha *et al.*, 2013) as described for their counterpart factors in SG assembly. Among the proteins that may drive PB assembly proteins are Rck/p54 (also known as DDX6), Dcp1a, Xrn1, LSm14A, Dcp2, GW182 and Ecd4 (Anderson and Kedersha, 2009; Jonas and Izaurralde, 2013; Kedersha and Anderson, 2007). Interestingly, PV infection results in cleavage/degradation of two of these potential PB assembly factors, Dcp1a and Xrn1 (Fig. 6.2). PV $3C^{pro}$ cleaves Dcp1a, likely at a C-terminal location that removes the terminal domain and phosphorylation site that mediates Dcp1a trimerization. In uninfected cells this trimerization stimulates efficient decapping of mRNA transcripts, the incorporation of Dcp1a into active decapping complexes with Dcp2 and Edc4 and also mediates Dcp1a localization to P bodies (Rzeczkowski *et al.*, 2011; Tritschler *et al.*, 2009). Xrn1 is not cleaved by PV 2A or 3C proteinases *in vitro*, however, PV infection resulted in destabilization of the Xrn1 half-life by 3-fold. Xrn1 destabilization was MG132-sensitive, indicating PV sharply stimulates proteasomal decay of Xrn1 (Dougherty *et al.*, 2011).

PV also targets the poly(A) nuclease subunit Pan3 through proteasome-independent degradation or cleavage (Fig. 6.2). Pan3 depletion blocks PB formation by inhibiting the first deadenylation step required for mRNPs to be included in these granules (Zheng *et al.*, 2008). Thus, cleavage of Pan3 likely provides a second mechanism to disrupt PBs through forced stabilization of host mRNA poly(A) tails. It is unknown at this time if Pan3 cleavage has an added benefit of stabilizing the length of viral poly(A) tails on mRNAs, since many mRNAs are subject to translation-associated poly(A) shortening (Laird-Offringa *et al.*, 1990, Grosset *et al.*, 2000). Presumably some viral mRNAs can initiate *de novo* RNA replication complexes, as happens with the infecting genomic RNA; thus, stabilizing the length of poly(A) tail may aid the reiterative transcription mechanism that also maintains poly(A) tail length on viral RNAs (Steil *et al.*, 2010). Although the methods employed by Enteroviruses to disassemble PBs remain uncertain, the mechanisms described above likely contribute to the observed effects. However, it is equally likely that additional virus-induced processes that help destabilize PBs will be discovered. A recent study that screened all PV viral proteins for destabilizing effects on PBs revealed that multiple viral proteins can reduce PB levels in cells. This included $3C^{pro}$, which can cleave Dcp1a as indicated above. However, $2A^{pro}$, 3CD and even 3Dpolymerase decreased PB levels when expressed in cells. In particular, $2A^{pro}$ strongly repressed PBs, however no molecular targets of $2A^{pro}$ that affect PB formation are known at this time (Dougherty *et al.*, 2015). In addition, the PB-promoting pathways that are disrupted by the two viral proteinases are different, suggesting unique molecular targets exist for each that regulate PBs. Oxidative stress from arsenite treatment is well known to drive PB formation, which could rescue PBs disrupted by $2A^{pro}$. However, PBs disrupted by $3C^{pro}$ could not be rescued by applying oxidative stress (Dougherty *et al.*, 2015).

Why Enteroviruses must antagonize RNA granules

Stress granules and P-bodies are an extension of translation control

Since cytoplasmic RNA granules are extensions of translation control mechanisms, viruses have a lot to lose by not regulating these bodies. First, Enteroviruses must achieve efficient and high levels of replication to stay in front of activating host defence mechanisms and cannot afford to have viral mRNA shuttled into translation-repression compartments. Second, even if the virus has a mechanism to ensure exclusion of viral mRNAs from any SGs that form, SG formation involving host mRNPs will sequester a large portion of translation factors and ribosome subunits away from easy utilization by the virus for translation of its own viral proteins. Again, the *Enterovirus* does not need any of the most important molecular tools to be locked in a closet. Similar arguments hold for PBs, which also repress the translation of their constituent mRNPs and may facilitate their decay. However, no experiments have been carried out to distinguish if the presence of P-bodies *per se* is detrimental to Enteroviruses, as opposed to P-body constituents that also play roles in PB assembly or the RNA-decay enzymes that are enriched in PBs. This is distinct from SG research where organized SGs can be induced and maintained in the presence of infection with expression of cleavage-resistant G3BP1 (Reineke and Lloyd, 2015; White *et al.*, 2007).

Stress granules may be platforms to signal innate immunity

An attractive hypothesis that is rapidly accumulating experimental evidence is that SGs can act as signalling platforms to amplify innate immunity. Thus, an additional important reason that viruses target stress granules is to control signalling. It is easy to imagine that infecting viruses have been causing all types of cellular stress in lower and higher eukaryotes for hundreds of millions of years, thus a linkage between stress sensing mechanisms and those that raise the innate immunity alarms makes sense. Much has been discovered about pathogen associated molecular patterns (PAMPs) that are recognized by Toll-like receptors (TLRs) and other pattern recognition receptors (PRRs) in both plants and animal systems. The specific interaction of a viral PAMP with a PRR or TLR results in activation of innate immune signalling through several pathways, stimulating both the interferon (IFN) and nuclear factor κB (NF-κB) arms of the innate immune system. However, these systems rely on pathogen recognition through detection of discrete and specific PAMPs on virus proteins or its RNA or DNA. This is where linkage between cell stress signals and innate immunity can provide key additional routes to sense infections since recognition and response to the various types of stress a pathogen may induce does not require a specific PAMP–PRR interaction. Such a scenario could heighten cell defences against invading pathogens in the absence of successful PAMP recognition, or could synergistically strengthen innate immune responses if PAMP recognition occurs. Further, reaction to a stress via SG formation is generally synonymous with pro-survival mechanisms. For instance, stress granules can protect from triggering apoptosis through inclusion of RACK1 or WDR62 into SG to control c-Jun N-terminal kinase (JNK) signalling (Arimoto *et al.*, 2008; Wasserman *et al.*, 2010).

There are now several links between stress granules and components of innate immunity. Many involve PKR, which is a classic interferon stimulated gene (ISG), but also is a canonical activator of stress granule formation through phosphorylation of its primary substrate eIF2α. PKR coordinates cellular stress, metabolic homeostasis and pathogen sensing and helps regulate JNK activation (Nakamura *et al.*, 2010). PKR activates inflammasomes in macrophages and innate immune transcriptional responses (García *et al.*, 2006; Lu *et al.*, 2012; Taghavi and Samuel, 2012). On the other hand, G3BP1 can drive assembly of stress granules in the absence of applied stress or infection. This causes activation of PKR and downstream eIF2-dependent translation inhibition (Reineke *et al.*, 2012). PKR enters virus-induced SGs in influenza-infected cells (Onomoto *et al.*, 2012), after stimulation of PKR with dsRNA (Zhang *et al.*, 2014) and after G3BP1 expression in the absence of infection (Reineke and Lloyd, 2015). This latter recruitment of PKR to stress granules during G3BP1 expression suggests a unique activation pathway exists that is dependent on SG formation involving G3BP1 (Reineke *et al.*, 2012).

Indeed new findings indicate that G3BP1 directly interacts with unactivated PKR and modulates its activation in a complex containing the G3BP1 interacting protein Caprin1. This PKR activation occurs in conjunction with its recruitment to SGs by G3BP1 (Reineke *et al.*, 2015). Once PKR becomes activated it can signal downstream to several innate immune effectors. NF-κB functions in many immune and inflammatory responses and can be activated by PKR, which interacts with the IκB kinase complex, promoting dissociation of the inhibitor IκBα from NF-κB (Bonnet *et al.*, 2000; Gil *et al.*, 2001). We have also found that G3BP1 can activate NF-κB through a direct binding interaction with its inhibitor IκBα. Interestingly, this G3BP1-IκBα interaction and subsequent activation of the NF-κB transcriptional cascade does not require interaction within SGs (R.E. Lloyd, unpublished). This indicates that G3BP1 plays multiple roles in innate immune activation both within and outside SGs.

G3BP and PKR can interact at a different level also. G3BP1, G3BP2 and Caprin1, which are all RNA-binding proteins, can also stimulate translation of some ISGs in response to IFN activation through unknown mechanisms. Interestingly, this includes stimulation of PKR mRNA translation itself (Bidet *et al.*, 2014).

Innate immune signalling factors concentrate in stress granules

Recent reports indicate that both upstream IFN activator PRRs and downstream ISGs can become concentrated within SGs. In addition to PKR, discussed above, three other important PRRs (RIG-I, MDA-5, LGP2) that sense viral RNA are concentrated within SGs after arsenite stress or after infection with a NS1 mutant influenza virus. A possible function of SG-based sentinel mechanism in IFN activation was proposed by observation of a loss of IFN-β mRNA production after depletion of PKR or G3BP1, the latter of which depletes SGs. Also, SGs did not form in PKR knockout MEFs during influenza infection, demonstrating PKR is upstream of SG induction in this system (Khaperskyy *et al.*, 2012; Onomoto *et al.*, 2012). However, it has been more difficult to directly implicate inclusion of innate immunity signalling factors within SGs with heightened activation of innate immunity. In the case of G3BP1-induced SG, comparison of expression of full-length G3BP1 with truncated G3BP1 that cannot form granules indicated that antiviral activity of G3BP1 was higher when SGs form (Reineke and Lloyd, 2015). One way that SG inclusion may drive PRR activation is through promoting interaction with the E3 ubiquitin ligase MEX3C. MEX3C causes the lysine-63-linked ubiquitination of RIG-I and downstream activation of the IFN-β promoter. This may be promoted within SGs since MEX3C and RIG-I concentrate and colocalize in the stress granules induced by Newcastle-Disease virus (Kuniyoshi *et al.*, 2014). Interestingly, the closely related paralogues MEX3A and MEX3B are recruited to PBs, though their function there is unclear (Buchet-Poyau *et al.*, 2007).

In addition to upstream PRRs, downstream ISGs are also found to concentrate in SGs. During infection with mutant influenza A virus, OAS and RNAse L were also reported in SGs (Onomoto *et al.*, 2012). OAS and RNAse L also are concentrated to variable extents in SGs induced in several cell lines by G3BP1- overexpression, in the absence of virus infection (Reineke and Lloyd, 2015). Though intriguing, it is less clear what roles might be played by inclusion of these downstream IFN response proteins within SGs. Since RNAse L degrades viral RNA, its activation may be somehow enhanced in SGs. However, any cleavage of enteroviral RNA would be require exit from SGs since viral RNA does not enter SGs (Piotrowska *et al.*, 2010). In contrast, virus RNA does enter the SG during infection with mutant influenza A viruses, thus inclusion of viral RNA in SGs is variable, depending on the type of virus, if WT versus mutant forms of virus are used, since inclusion is likely counteracted by viral proteins produced by WT virus. This illustrates a battle that viruses must control and may lose during unsuccessful infections.

So how does all this stack up for Enteroviruses? *Enterovirus* proteinases have evolved into a niche of substrate specificity that can cleave many key innate immune signalling proteins, including MDA5, RIG-I, MAVS, IRF7 and TRIF (Barral *et al.*, 2007, 2009; Feng *et al.*, 2014; Lei *et al.*, 2011, 2013; Mukherjee *et al.*, 2011). So with this arsenal of innate immune attack strategies, what further impact can G3BP1 cleavage provide? Early investigation revealed that blockage of G3BP1 cleavage by expression of a cleavage-resistant mutant of G3BP1 resulted in a 7-fold drop in virus titre. Experiments from adding G3BP1 back into G3BP−/− MEFs, suggest the anti-viral impact can be even greater, resulting in 1–2 logs inhibition of CVB3, or EV70 replication (Reineke and Lloyd, 2015). Thus, cleavage of G3BP may be as important for blocking innate immune activating function of PKR as it is for blocking SG formation per se, killing two birds with one stone. It seems logical that concentration of innate immune factors in SGs may amplify signalling roles in innate immune activation, but this may not always be the case. The concentration of some innate immune factors in SGs may function to control infections with DNA viruses or negative strand RNA viruses, or play no role in controlling Enteroviruses. For example, MDA5 is more important than RIG-I for sensing Enteroviruses and both enter SGs (Feng *et al.*, 2012), but the role of SG entry for MDA5 signalling and interferon activation has been challenged. Infection with a mutant Mengo virus that has an inactive leader protein that cannot repress IFN-α/β activation, induces SGs in a PKR-dependent manner (Langereis *et al.*, 2013). In this system cells that were defective in SG formation yielded higher virus replication. This supports the notion that SGs play a role in antiviral defence. However, even though MDA5 entered SG, MDA5 activation did not occur and by blocking SG formation, IFN responses could still be activated. Thus, inclusion of MDA5 in SGs has no apparent role in activation of IFN this system (Langereis *et al.*, 2013) and mere inclusion of signalling molecules in SGs does not necessarily indicate their activation will occur. The results also do not exclude the possibility that the mutant *Mengo virus* can limit MDA5 activation via other mechanisms. On the other hand, the entry of PRRs into SGs may play roles during infection with other types of viruses. The functional impact of MDA5 entry into SGs in *Enterovirus* infection has not been reported.

Innate immune signalling through P-body constituents

Since P-bodies also are repositories of silenced transcripts that accumulate during stress, it is also possible that these bodies or their associated factors play roles in pathogen surveillance and innate immune activation. It is interesting to note that PKR enters PBs during human papilloma virus infection, though this may be associated with viral misdirection of PKR activation (Hebner *et al.*, 2006). In the case of Enteroviruses there is crosstalk between the PB factor Dcp1a and PKR activation that can block virus replication. Poliovirus 3C^{pro} cleaves Dcp1a at an unmapped site (Dougherty *et al.*, 2011). Though this cleavage may interfere with PB assembly as noted above, it is a curious target for viral aggression since Dcp1a is a regulator of mRNA decapping, and Enteroviruses do not have capped viral RNAs. Why does PV target this factor? We performed studies to map the 3C^{pro}-cleavage site of Dcp1a and found that while endogenous Dcp1a was cleaved by virus, the ectopic-expressed Dcp1a was not cleaved. Further investigation of this incongruity revealed that expression of Dcp1a prior to infection surprisingly restricts poliovirus infection. Dcp1a expression results in activation of PKR and subsequent phosphorylation of eIF2α, resulting in potent translation inhibition that restricted virus replication (Dougherty *et al.*, 2014). PKR activation requires the N-terminal EVH1 domain of Dcp1a. The translation blockade induced by Dcp1a expression suggests a novel signalling axis exists linking RNA degradation/decapping and regulation of translation. However as discussed above, PKR activation also initiates a cascade of downstream innate immune activation events. In this case, PKR activation did not require PB formation so the P-body per se may be less important than some of its constituents as a threat to efficient virus replication (Dougherty *et al.*, 2014). It will be interesting to see if additional links between PBs and innate immunity are revealed in the future.

Future directions

The field of RNA granules biology is relatively new and the subfield of RNA granule–virus interaction is even more recent. Accordingly, a large amount of work remains to be done, and many important questions remain unanswered. Focusing on Enteroviruses and their relationship with RNA granules, several key questions loom at the forefront and require investigation.

First, are there additional host factors beyond G3BP1, Dcp1a, Xrn1 and Pan3 that *Enteroviruses* target to control RNA granules? What are the unknown targets for 2A^{pro} that help control SG and PB assembly and how do those factors support SG or PB condensation?

Second, further study of the known viral targets is required to determine how they function in SG and PB assembly. How does viral cleavage of G3BP1 really block stress granule assembly? There is growing evidence that SGs form as a result of post-translational changes in key RNA binding proteins. For instance, G3BP1 is phosphorylated by casein kinase 2 to repress SG formation (Reineke *et al.*, 2017), and similarly, is methylated by protein arginine methyltransferase 1 and protein arginine methyltransferase 5, also to repress SG formation. In fact, protein arginine methyl marks are rapidly removed during cell stress to facilitate SG formation (Tsai *et al.*, 2016). This raises interesting questions whether protein methylation plays any role in protein interactions that regulate innate immune activation.

Third, there is growing evidence of linkages of SGs and PBs to mitochondria and energy/apoptosis control and autophagy that needs exploration. Electron microscopy has suggested that G3BP1-positive SGs localize near mitochondrial surfaces (Fung *et al.*, 2013). P-bodies also transiently interact with mitochondria in microtubule-dependent mechanism. While mitochondrial function was not dependent on PB contact, mitochondrial function did support PB-associated miRNA-mediated RNAi efficiency (Huang *et al.*, 2011). Further, as mentioned above, autophagy is also linked to turnover of SGs in yeast and mammalian cells (Buchan *et al.*, 2013; Seguin *et al.*, 2014). These findings all suggest that RNA granules are involved in communication with cell homeostasis and energy production systems. Since *Enteroviruses* are well known to regulate and hijack the autophagy machinery, determination of the linkage of these systems may impact understanding of cell biology at a systems level, in addition to providing more insights into virus–host interactions.

Finally, since control of innate immunity is of paramount importance in virology, determination of the extent of interactions of RNA granules with the innate immune system is critical. Details of how some of these interactions function at the mechanistic level are beginning to emerge. Much more work is required to unravel these complex interactions, but the promise is a much fuller and revised understanding of innate immune regulation as a whole.

Acknowledgements

This work was funded by NIH Public Health Service grants AI50237 (R.E.L.) and supported by the Integrated Microscopy Core at Baylor College of Medicine with funding from the NIH (HD007495, DK56338, and CA125123), the Dan L. Duncan Cancer Center, and the John S. Dunn Gulf Coast Consortium for Chemical Genomics.

References

Anderson, P., and Kedersha, N. (2009). RNA granules: post-transcriptional and epigenetic modulators of gene expression. Nat. Rev. Mol. Cell Biol. *10*, 430–436. https://doi.org/10.1038/nrm2694

Arimoto, K., Fukuda, H., Imajoh-Ohmi, S., Saito, H., and Takekawa, M. (2008). Formation of stress granules inhibits apoptosis by suppressing stress-responsive MAPK pathways. Nat. Cell Biol. *10*, 1324–1332. https://doi.org/10.1038/ncb1791

Aulas, A., Stabile, S., and Vande Velde, C. (2012). Endogenous TDP-43, but not FUS, contributes to stress granule assembly via G3BP. Mol. Neurodegener.7, 54.

Barral, P.M., Morrison, J.M., Drahos, J., Gupta, P., Sarkar, D., Fisher, P.B., and Racaniello, V.R. (2007). MDA-5 is cleaved in poliovirus-infected cells. J. Virol. *81*, 3677–3684.

Barral, P.M., Sarkar, D., Fisher, P.B., and Racaniello, V.R. (2009). RIG-I is cleaved during picornavirus infection. Virology *391*, 171–176. https://doi.org/10.1016/j.virol.2009.06.045

Bidet, K., Dadlani, D., and Garcia-Blanco, M.A. (2014). G3BP1, G3BP2 and CAPRIN1 Are Required for Translation of Interferon Stimulated mRNAs and Are Targeted by a Dengue Virus Non-coding RNA. PLOS Pathogens *10*, e1004242.

Black, T.L., Barber, G.N., and Katze, M.G. (1993). Degradation of the interferon-induced 68,000-M(r) protein kinase by poliovirus requires RNA. J. Virol. *67*, 791–800.

Bonnet, M.C., Weil, R., Dam, E., Hovanessian, A.G., and Meurs, E.F. (2000). PKR stimulates NF-kappaB irrespective of its kinase function by interacting with the IkappaB kinase complex. Mol. Cell. Biol. *20*, 4532–4542.

Borghese, F., and Michiels, T. (2011). The leader protein of cardioviruses inhibits stress granule assembly. J. Virol. *85*, 9614–9622. https://doi.org/10.1128/JVI.00480-11

Buchan, J.R., Kolaitis, R.M., Taylor, J.P., and Parker, R. (2013). Eukaryotic stress granules are cleared by autophagy and Cdc48/VCP function. Cell *153*, 1461–1474. https://doi.org/10.1016/j.cell.2013.05.037

Buchet-Poyau, K., Courchet, J., Le Hir, H., Séraphin, B., Scoazec, J.-Y., Duret, L., Domon-Dell, C., Freund, J.-N., and Billaud, M. (2007). Identification and characterization of human Mex-3 proteins, a novel family of evolutionarily conserved RNA-binding proteins differentially localized to processing bodies. Nucleic Acids Res. 35, 1289–1300.

Cathcart, A.L., Rozovics, J.M., and Semler, B.L. (2013). Cellular mRNA decay protein AUF1 negatively regulates enterovirus and human rhinovirus infections. J. Virol. *87*, 10423–10434. https://doi.org/10.1128/JVI.01049-13

Chernov, K.G., Barbet, A., Hamon, L., Ovchinnikov, L.P., Curmi, P.A., and Pastré, D. (2009). Role of microtubules in stress granule assembly: microtubule dynamical instability favors the formation of micrometric stress granules in cells. J. Biol. Chem. *284*, 36569–36580.

Dang, Y., Kedersha, N., Low, W.-K., Romo, D., Gorospe, M., Kaufman, R., Anderson, P., and Liu, J.O. (2006). Eukaryotic initiation factor 2alpha-independent pathway of stress granule induction by the natural product pateamine A. J. Biol. Chem. *281*, 32870–32878.

Dougherty, J.D., Reineke, L.C., and Lloyd, R.E. (2014). mRNA decapping enzyme 1a (Dcp1a)-induced translational arrest through protein kinase R (PKR) activation requires the N-terminal enabled vasodilator-stimulated protein homology 1 (EVH1) domain. J. Biol. Chem. *289*, 3936–3949.

Dougherty, J.D., Tsai, W.C., and Lloyd, R.E. (2015). Multiple Poliovirus Proteins Repress Cytoplasmic RNA Granules. Viruses 7, 6127–6140. https://doi.org/10.3390/v7122922

Dougherty, J.D., White, J.P., and Lloyd, R.E. (2011). Poliovirus-mediated disruption of cytoplasmic processing bodies. J. Virol. *85*, 64–75. https://doi.org/10.1128/JVI.01657-10

Emara, M.M., Fujimura, K., Sciaranghella, D., Ivanova, V., Ivanov, P., and Anderson, P. (2012). Hydrogen peroxide induces stress granule formation independent of eIF2α phosphorylation. Biochem. Biophys. Res. Commun. *423*, 763–769.

Etchison, D., Milburn, S.C., Edery, I., Sonenberg, N., and Hershey, J.W. (1982). Inhibition of HeLa cell protein synthesis following poliovirus infection correlates with the proteolysis of a 220,000-dalton polypeptide associated with eucaryotic initiation factor 3 and a cap binding protein complex. J. Biol. Chem. *257*, 14806–14810.

Feng, Q., Hato, S.V., Langereis, M.A., Zoll, J., Virgen-Slane, R., Peisley, A., Hur, S., Semler, B.L., van Rij, R.P., and van Kuppeveld, F.J. (2012). MDA5 detects the double-stranded RNA replicative form in picornavirus-infected cells. Cell Rep 2, 1187–1196. https://doi.org/10.1016/j.celrep.2012.10.005

Feng, Q., Langereis, M.A., Lork, M., Nguyen, M., Hato, S.V., Lanke, K., Emdad, L., Bhoopathi, P., Fisher, P.B., Lloyd, R.E., *et al.* (2014). Enterovirus 2Apro targets MDA5 and MAVS in infected cells. J. Virol. *88*, 3369–3378. https://doi.org/10.1128/JVI.02712-13

Fung, G., Ng, C.S., Zhang, J., Shi, J., Wong, J., Piesik, P., Han, L., Chu, F., Jagdeo, J., Jan, E., *et al.* (2013). Production of a Dominant-Negative Fragment Due to G3BP1 Cleavage Contributes to the Disruption of Mitochondria-Associated Protective Stress Granules during CVB3 Infection. PLOS ONE *8*, e79546.

García, M.A., Gil, J., Ventoso, I., Guerra, S., Domingo, E., Rivas, C., and Esteban, M. (2006). Impact of protein kinase PKR in cell biology: from antiviral to antiproliferative action. Microbiol. Mol. Biol. Rev. *70*, 1032–1060.

Gil, J., Rullas, J., García, M.A., Alcamí, J., and Esteban, M. (2001). The catalytic activity of dsRNA-dependent protein kinase, PKR, is required for NF-kappaB activation. Oncogene *20*, 385–394. https://doi.org/10.1038/sj.onc.1204109

Gilks, N., Kedersha, N., Ayodele, M., Shen, L., Stoecklin, G., Dember, L.M., and Anderson, P. (2004). Stress granule assembly is mediated by prion-like aggregation of TIA-1. Mol. Biol. Cell *15*, 5383–5398. https://doi.org/10.1091/mbc.E04-08-0715

Gradi, A., Svitkin, Y.V., Imataka, H., and Sonenberg, N. (1998). Proteolysis of human eukaryotic translation initiation factor eIF4GII, but not eIF4GI, coincides with the shutoff of host protein synthesis after poliovirus infection. Proc Natl Acad Sci USA *95*, 11089–11094.

Grosset, C., Chen, C.-Y., Xu, N., Sonenberg, N., Jacquemin-Sablon, H., and Shyu, A.-B. (2000). A mechanism for translationally coupled mRNA turnover: interaction between the poly(A) tail and a c-fos RNA coding determinant via a protein complex. Cell *103*, 29–40.

Hebner, C.M., Wilson, R., Rader, J., Bidder, M., and Laimins, L.A. (2006). Human papillomaviruses target the double-stranded RNA protein kinase pathway. J. Gen. Virol. *87*, 3183–3193.

Huang, L., Mollet, S., Souquere, S., Le Roy, F., Ernoult-Lange, M., Pierron, G., Dautry, F., and Weil, D. (2011). Mitochondria associate with P-bodies and modulate microRNA-mediated RNA interference. J. Biol. Chem. *286*, 24219–24230. https://doi.org/10.1074/jbc.M111.240259

Jonas, S., and Izaurralde, E. (2013). The role of disordered protein regions in the assembly of decapping complexes and RNP granules. Genes Dev. *27*, 2628–2641. https://doi.org/10.1101/gad.227843.113

Kedersha, N., and Anderson, P. (2007). Mammalian stress granules and processing bodies. Meth. Enzymol. *431*, 61–81.

Kedersha, N., Ivanov, P., and Anderson, P. (2013). Stress granules and cell signaling: more than just a passing phase? Trends Biochem. Sci. *38*, 494–506. https://doi.org/10.1016/j.tibs.2013.07.004

Kedersha, N., Stoecklin, G., Ayodele, M., Yacono, P., Lykke-Andersen, J., Fritzler, M.J., Scheuner, D., Kaufman, R.J., Golan, D.E., and Anderson, P. (2005). Stress granules and processing bodies are dynamically linked sites of mRNP remodeling. J. Cell Biol. *169*, 871–884.

Khaperskyy, D.A., Hatchette, T.F., and McCormick, C. (2012). Influenza A virus inhibits cytoplasmic stress granule formation. FASEB J. *26*, 1629–1639. https://doi.org/10.1096/fj.11-196915

Khong, A., and Jan, E. (2011). Modulation of stress granules and P bodies during dicistrovirus infection. J. Virol. *85*, 1439–1451. https://doi.org/10.1128/JVI.02220-10

Khong,A., Kerr, C.H., Yeung, C.H.L., Keatings, K., Nayak, A., Allan, D.W., and Jan, E. (2016). Disruption of stress granule formation by the multifunctional cricket paralysis virus 1A protein. J. Virol. epub 21 Dec. https://doi.org/10.1128/JVI.01779-16

Kobayashi, M., Wilson, A.C., Chao, M.V., and Mohr, I. (2012). Control of viral latency in neurons by axonal mTOR signaling and the 4E-BP translation repressor. Genes Dev. *26*, 1527–1532. https://doi.org/10.1101/gad.190157.112

Kuniyoshi, K., Takeuchi, O., Pandey, S., Satoh, T., Iwasaki, H., Akira, S., and Kawai, T. (2014). Pivotal role of RNA-binding E3 ubiquitin ligase MEX3C in RIG-I-mediated antiviral innate immunity. Proc. Natl. Acad. Sci. U.S.A. *111*, 5646–5651. https://doi.org/10.1073/pnas.1401674111

Kuyumcu-Martinez, N.M., Van Eden, M.E., Younan, P., and Lloyd, R.E. (2004). Cleavage of poly(A)-binding protein by poliovirus 3C protease inhibits host cell translation: a novel mechanism for host translation shutoff. Mol. Cell. Biol. *24*, 1779–1790.

Kwon, S., Zhang, Y., and Matthias, P. (2007). The deacetylase HDAC6 is a novel critical component of stress granules involved in the stress response. Genes Dev. *21*, 3381–3394.

Laird-Offringa, I.A., de Wit, C.L., Elfferich, P., and van der Eb, A.J. (1990). Poly(A) tail shortening is the translation-dependent step in c-myc mRNA degradation. Mol. Cell. Biol. *10*, 6132–6140.

Langereis, M.A., Feng, Q., and van Kuppeveld, F.J. (2013). MDA5 localizes to stress granules, but this localization is not required for the induction of type I interferon. J. Virol. *87*, 6314–6325. https://doi.org/10.1128/JVI.03213-12

Langereis, M.A., Feng, Q., Nelissen, F.H., Virgen-Slane, R., van der Heden van Noort, G.J., Maciejewski, S., Filippov, D.V., Semler, B.L., van Delft, F.L., and van Kuppeveld, F.J. (2014). Modification of picornavirus genomic RNA using 'click' chemistry shows that unlinking of the VPg peptide is dispensable for translation and replication of the incoming viral RNA. Nucleic Acids Res. *42*, 2473–2482. https://doi.org/10.1093/nar/gkt1162

Lei, X., Sun, Z., Liu, X., Jin, Q., He, B., and Wang, J. (2011). Cleavage of the adaptor protein TRIF by enterovirus 71 3C inhibits antiviral responses mediated by Toll-like receptor 3. J. Virol. *85*, 8811–8818. https://doi.org/10.1128/JVI.00447-11

Lei, X., Xiao, X., Xue, Q., Jin, Q., He, B., and Wang, J. (2013). Cleavage of interferon regulatory factor 7 by enterovirus 71 3C suppresses cellular responses. J. Virol. *87*, 1690–1698. https://doi.org/10.1128/JVI.01855-12

Lloyd, R.E. (2012). How do viruses interact with stress-associated RNA granules? PLOS Pathog. *8*, e1002741. https://doi.org/10.1371/journal.ppat.1002741

Loschi, M., Leishman, C.C., Berardone, N., and Boccaccio, G.L. (2009). Dynein and kinesin regulate stress-granule and P-body dynamics. J. Cell Sci. *122*, 3973–3982.

Lu, B., Nakamura, T., Inouye, K., Li, J., Tang, Y., Lundbäck, P., Valdes-Ferrer, S.I., Olofsson, P.S., Kalb, T., Roth, J., *et al.* (2012). Novel role of PKR in inflammasome activation and HMGB1 release. Nature *488*, 670–674. https://doi.org/10.1038/nature11290

Matsuki, H., Takahashi, M., Higuchi, M., Makokha, G.N., Oie, M., and Fujii, M. (2013). Both G3BP1 and G3BP2 contribute to stress granule formation. Genes Cells *18*, 135–146. https://doi.org/10.1111/gtc.12023

Mazroui, R., Sukarieh, R., Bordeleau, M.-E., Kaufman, R.J., Northcote, P., Tanaka, J., Gallouzi, I., and Pelletier, J. (2006). Inhibition of ribosome recruitment induces stress granule formation independently of eukaryotic initiation factor 2alpha phosphorylation. Mol. Biol. Cell. *17*, 4212–4219.

Mukherjee, A., Morosky, S.A., Delorme-Axford, E., Dybdahl-Sissoko, N., Oberste, M.S., Wang, T., and Coyne, C.B. (2011). The coxsackievirus B 3C protease cleaves MAVS and TRIF to attenuate host type I interferon and apoptotic signaling. PLOS Pathog. *7*, e1001311. https://doi.org/10.1371/journal.ppat.1001311

Nadezhdina, E.S., Lomakin, A.J., Shpilman, A.A., Chudinova, E.M., and Ivanov, P.A. (2010). Microtubules govern stress granule mobility and dynamics. Biochim. Biophys. Acta *1803*, 361–371. https://doi.org/10.1016/j.bbamcr.2009.12.004

Nakamura, T., Furuhashi, M., Li, P., Cao, H., Tuncman, G., Sonenberg, N., Gorgun, C.Z., and Hotamisligil, G.S. (2010). Double-stranded RNA-dependent protein kinase links pathogen sensing with stress and metabolic homeostasis. Cell *140*, 338–348.

Ng, C.S., Jogi, M., Yoo, J.-S., Onomoto, K., Koike, S., Iwasaki, T., Yoneyama, M., Kato, H., and Fujita, T. (2013). Encephalomyocarditis virus disrupts stress granules, the critical platform for triggering antiviral innate immune responses. J. Virol. *87*, 9511–9522.

O'Neill, R.E., and Racaniello, V.R. (1989). Inhibition of translation in cells infected with a poliovirus 2Apro mutant correlates with phosphorylation of the alpha subunit of eucaryotic initiation factor 2. J. Virol. *63*, 5069–5075.

Ohn, T., Kedersha, N., Hickman, T., Tisdale, S., and Anderson, P. (2008). A functional RNAi screen links O-GlcNAc modification of ribosomal proteins to stress granule and processing body assembly. Nat. Cell Biol. *10*, 1224–1231. https://doi.org/10.1038/ncb1783

Onomoto, K., Jogi, M., Yoo, J.S., Narita, R., Morimoto, S., Takemura, A., Sambhara, S., Kawaguchi, A., Osari, S., Nagata, K., *et al.* (2012). Critical role of an antiviral stress granule containing RIG-I and PKR in viral detection and innate immunity. PLOS ONE 7, e43031. https://doi.org/10.1371/journal.pone.0043031

Piotrowska, J., Hansen, S.J., Park, N., Jamka, K., Sarnow, P., and Gustin, K.E. (2010). Stable formation of compositionally unique stress granules in virus-infected cells. J. Virol. *84*, 3654–3665. https://doi.org/10.1128/JVI.01320-09

Reineke, L.C., and Lloyd, R.E. (2013). Diversion of stress granules and P-bodies during viral infection. Virology *436*, 255–267. https://doi.org/10.1016/j.virol.2012.11.017

Reineke, L.C., and Lloyd, R.E. (2015). The stress granule protein G3BP1 recruits protein kinase R To promote multiple innate immune antiviral responses. J. Virol. *89*, 2575–2589.

Reineke, L.C., Dougherty, J.D., Pierre, P., and Lloyd, R.E. (2012). Large G3BP-induced granules trigger eIF2α phosphorylation. Mol. Biol. Cell *23*, 3499–3510. https://doi.org/10.1091/mbc.E12-05-0385

Reineke, L.C., Kedersha, N., Langereis, M.A., van Kuppeveld, F.J.M., and Lloyd, R.E. (2015). Stress granules regulate double-stranded RNA-dependent protein kinase activation through a complex containing G3BP1 and Caprin1. MBio *6*, e02486.

Reineke, L.C., Tsai, W.C., Jain, A., Kaelber, J.T. Sung S.Y., and R.E. Lloyd (2017). Casein Kinase 2 regulates stress granule dynamics by phosphorylating the stress granule nucleating protein G3BP1. Mol. Cell. Biol. *37*, e00596-16.

Rozovics, J.M., Chase, A.J., Cathcart, A.L., Chou, W., Gershon, P.D., Palusa, S., Wilusz, J., and Semler, B.L. (2012). Picornavirus modification of a host mRNA decay protein. MBio *3*, e00431–12. https://doi.org/10.1128/mBio.00431-12

Rzeczkowski, K., Beuerlein, K., Müller, H., Dittrich-Breiholz, O., Schneider, H., Kettner-Buhrow, D., Holtmann, H., and Kracht, M. (2011). c-Jun N-terminal kinase phosphorylates DCP1a to control formation of P bodies. J. Cell Biol. *194*, 581–596. https://doi.org/10.1083/jcb.201006089

Seguin, S.J., Morelli, F.F., Vinet, J., Amore, D., De Biasi, S., Poletti, A., Rubinsztein, D.C., and Carra, S. (2014). Inhibition of autophagy, lysosome and VCP function impairs stress granule assembly. Cell Death Differ. *21*, 1838–1851. https://doi.org/10.1038/cdd.2014.103

Steil, B.P., Kempf, B.J., and Barton, D.J. (2010). Poly(A) at the 3' end of positive-strand RNA and VPg-linked poly(U) at the 5' end of negative-strand RNA are reciprocal templates during replication of poliovirus RNA. J. Virol. *84*, 2843–2858. https://doi.org/10.1128/JVI.02620-08

Taghavi, N., and Samuel, C.E. (2012). Protein kinase PKR catalytic activity is required for the PKR-dependent activation of mitogen-activated protein kinases and amplification of interferon beta induction following virus infection. Virology *427*, 208–216.

Teixeira, D., Sheth, U., Valencia-Sanchez, M.A., Brengues, M., and Parker, R. (2005). Processing bodies require RNA for assembly and contain nontranslating mRNAs. RNA *11*, 371–382. Anderson, P., and Kedersha, N. (2009). RNA granules: post-transcriptional and epigenetic modulators of gene expression. Nat. Rev. Mol. Cell Biol. *10*, 430–436. https://doi.org/10.1038/nrm2694

Tourrière, H., Chebli, K., Zekri, L., Courselaud, B., Blanchard, J.M., Bertrand, E., and Tazi, J. (2003). The RasGAP-associated endoribonuclease G3BP assembles stress granules. J. Cell Biol. *160*, 823–831. https://doi.org/10.1083/jcb.200212128

Tritschler, F., Braun, J.E., Motz, C., Igreja, C., Haas, G., Truffault, V., Izaurralde, E., and Weichenrieder, O. (2009). DCP1 forms asymmetric trimers to assemble into active mRNA decapping complexes in metazoa. Proc. Natl. Acad. Sci. U.S.A. *106*, 21591–21596. https://doi.org/10.1073/pnas.0909871106

Tsai, N.P., Tsui, Y.C., and Wei, L.N. (2009). Dynein motor contributes to stress granule dynamics in primary neurons. Neuroscience *159*, 647–656. https://doi.org/10.1016/j.neuroscience.2008.12.053

Tsai, W.C., and Lloyd, R.E. (2014). Cytoplasmic RNA Granules and Viral Infection. Annu Rev Virol *1*, 147–170. https://doi.org/10.1146/annurev-virology-031413-085505

Tsai, W.C., Gayatri, S., Reineke, L.C., Sbardella, G., Bedford, M.T., and Lloyd, R.E. (2016). Arginine Demethylation of G3BP1 Promotes Stress Granule Assembly. J. Biol. Chem. *291*, 22671–22685.

Virgen-Slane, R., Rozovics, J.M., Fitzgerald, K.D., Ngo, T., Chou, W., van der Heden van Noort, G.J., Filippov, D.V., Gershon, P.D., and Semler, B.L. (2012). An RNA virus hijacks an incognito function of a DNA repair enzyme. Proc. Natl. Acad. Sci. U.S.A. *109*, 14634–14639. https://doi.org/10.1073/pnas.1208096109

Wasserman, T., Katsenelson, K., Daniliuc, S., Hasin, T., Choder, M., and Aronheim, A. (2010). A novel c-Jun N-terminal kinase (JNK)-binding protein WDR62 is recruited to stress granules and mediates a nonclassical JNK activation. Mol. Biol. Cell *21*, 117–130. https://doi.org/10.1091/mbc.E09-06-0512

White, J.P., and Lloyd, R.E. (2011). Poliovirus unlinks TIA1 aggregation and mRNA stress granule formation. J. Virol. *85*, 12442–12454. https://doi.org/10.1128/JVI.05888-11

White, J.P., Cardenas, A.M., Marissen, W.E., and Lloyd, R.E. (2007). Inhibition of cytoplasmic mRNA stress granule formation by a viral proteinase. Cell Host Microbe *2*, 295–305.

White, J.P., Reineke, L.C., and Lloyd, R.E. (2011). Poliovirus switches to an eIF2-independent mode of translation during infection. J. Virol. *85*, 8884–8893. https://doi.org/10.1128/JVI.00792-11

Wong, J., Si, X., Angeles, A., Zhang, J., Shi, J., Fung, G., Jagdeo, J., Wang, T., Zhong, Z., Jan, E., *et al.* (2013). Cytoplasmic redistribution and cleavage of AUF1 during coxsackievirus infection enhance the stability of its viral genome. Faseb J. *27*, 2777-2787.

Wu, S., Lin, L., Zhao, W., Li, X., Wang, Y., Si, X., Wang, T., Wu, H., Zhai, X., Zhong, X., *et al.* (2014a). AUF1 is recruited to the stress granules induced by coxsackievirus B3. Virus Res. *192*, 52–61. https://doi.org/10.1016/j.virusres.2014.08.003

Wu, S., Wang, Y., Lin, L., Si, X., Wang, T., Zhong, X., Tong, L., Luan, Y., Chen, Y., Li, X., *et al.* (2014b). Protease 2A induces stress granule formation during coxsackievirus B3 and enterovirus 71 infections. Virol. J. *11*, 192. https://doi.org/10.1186/s12985-014-0192-1

Xie, W., and Denman, R.B. (2011). Protein methylation and stress granules: posttranslational remodeler or innocent bystander? Mol. Biol. International *2011*, 137459.

Zamora, M., Marissen, W.E., and Lloyd, R.E. (2002). Multiple eIF4GI-specific protease activities present in uninfected and poliovirus-infected cells. J. Virol. *76*, 165–177.

Zhang, P., Li, Y., Xia, J., He, J., Pu, J., Xie, J., Wu, S., Feng, L., Huang, X., and Zhang, P. (2014). IPS-1 plays an essential role in dsRNA-induced stress granule formation by interacting with PKR and promoting its activation. J. Cell Sci. *127*, 2471–2482.

Zheng, D., Ezzeddine, N., Chen, C.Y., Zhu, W., He, X., and Shyu, A.B. (2008). Deadenylation is prerequisite for P-body formation and mRNA decay in mammalian cells. J. Cell Biol. *182*, 89–101. https://doi.org/10.1083/jcb.200801196

The Autophagic Pathway and *Enterovirus* Infection

7

William T. Jackson

Department of Microbiology and Immunology, University of Maryland School of Medicine, Baltimore, MD, USA.

Correspondence: wjackson@som.umaryland.edu

https://doi.org/10.21775/9781910190739.07

Abstract

A common feature of the genus *Enterovirus* is the rapid rearrangement of the cytoplasm of infected cells. In general, cellular organelles are altered or disrupted and virus-specific vesicles, which take up a large perinuclear region of the cytoplasm, proliferate. One role of the newly generated vesicles is to act as substrates for RNA-dependent RNA replication complexes. One subset of these vesicles, observed later in infection, is a double-membraned population which strongly resemble autophagosomes, the organelles of the autophagy pathway. Autophagy, a degradative mechanism which maintains cellular homeostasis, is markedly increased during times of cellular stress, including starvation and infection. Autophagy is an important part of the innate immune response, degrading cytosolic pathogens and providing peptides for MHC presentation. However, for a subset of pathogens, including many *Enteroviruses*, the autophagic pathway is subverted to promote replication of the invader. In this chapter we discuss the ways in which *Enteroviruses* are known to interact with the autophagy pathway, often to promote viral replication.

Autophagy was first identified in *Enterovirus* infections through electron microscopy

In 1959 electron microscopic analysis of poliovirus-infected, lysed HeLa cells revealed the presence of vesicles, some of which appeared to contain virus (Horne and Nagington, 1959). In 1965, Palade and colleagues performed electron microscopy on intact, poliovirus-infected HeLa cells and observed double-membraned vesicles, frequently containing virus, proliferating throughout the cytoplasm (Dales *et al.*, 1965). In these images, the cytoplasm of the cells is filled with vesicles resembling donuts and crescents. Each of these contains two lipid bilayers separated by an electron-light region, and the interior resembles cytoplasm. Some of the structures appear to be open to the cytoplasm. Near and within these vesicles are apparent virions. The authors suggested these represent 'a secondary response to infection,' although, as they pointed out, this extreme response does little good, as the cell is doomed to lyse a short time later regardless of the response (Dales *et al.*, 1965).

These vesicles were observed late in infection, at 7 hours post infection (hpi), although subsequent experiments showed that these vesicles were present as early as 5 hpi and evidence of autophagic signalling present as early as 3 hpi (Suhy *et al.*, 2000; Taylor and Kirkegaard, 2007).

Autophagosome-like vesicles have been observed during coxsackievirus infection *in vivo* as well. Electron microscopy of CVB3-infected newborn mouse pancreas displays both multilamellar structures and virus bound by single membranes (Harrison *et al.*, 1972). Studies using coxsackievirus B4 have identified similar structures, including spherical aggregates of virions and membrane–virion complexes (MVCs) (Harb and Burch, 1975). It was recently shown that CVB3 can also induce extremely large structures, termed 'megaphagosomes,' in pancreatic cells. Some of these megaphagosomes are so large they can be measured in micrometres (Kemball *et al.*, 2010). The presence of these multilamellar structures indicate that autophagic regulation (or dysregulation) and membrane remodelling are likely taking place in the infected cells, although they do not necessarily indicate genuine autophagy, as will be explained in the next section.

Modern imaging has brought more precise structural understanding of vesicular development during infection. In 1996 the Kirkegaard group, using high-pressure freezing electron microscopy, reported on the formation of double-membraned autophagosome-like structures in poliovirus-infected cells (Schlegel *et al.*, 1996). However, where typical autophagosomes are 500 μM in diameter or greater, the PV-induced structures appeared to be 200–500 μM across. The Ehrenfeld group demonstrated the interconnected nature of poliovirus replication membranes using 3D-electron tomography. Early in infection, the membranes induced by poliovirus resemble a series of highly interconnected tubes. Later in infection, the membranes appear more individual, with several images suggestive of vesicles budding or pinching from tubes. Finally, by 5–6 hours post infection, double-membraned vesicles predominate. The authors concluded that the replication membranes are something akin to the convoluted membranes or membranous web identified in studies of flavivirus replication (Belov *et al.*, 2012). The authors went on to suggest that the early complex structures are progenitors of the later double-membraned vesicles.

Double-membraned vesicles and other indications of autophagy have been found in multiple *Enteroviruses*, including coxsackievirus B3 and B4 (CVB3, CVB4), rhinovirus 2 (HRV-2), foot-and-mouth-disease virus (FMDV) encephalomyocarditis virus (EMCV) and enterovirus 71 (EV71) (Huang *et al.*, 2009; Klein and Jackson, 2011; O'Donnell *et al.*, 2011; Wong *et al.*, 2008; Yoon *et al.*, 2008; Zhang *et al.*, 2011) It is worth noting here that two *Enteroviruses*, HRV-1A and Hepititis A Virus (HAV), do not appear to engage or benefit from the autophagic pathway (Feng *et al.*, 2013; Quiner and Jackson, 2010). For the majority, however, induction and subversion of autophagy appears to be important for a normal viral infection cycle.

Basics of autophagy

Autophagy is a constitutive cellular pathway which enables the bulk degradation of cytosolic contents, including organelles, to maintain cellular turnover and homeostasis (González-Polo *et al.*, 2016). During times of cellular stress, including infection, starvation, or stages of embryonic development, autophagy is up-regulated to provide an emergency source of amino acids or to clear the cytoplasm. There are multiple defined variants of autophagy, including specific degradation and turnover of mitochondria (mitophagy), degradation of foreign invaders (xenophagy) and chaperone-mediated autophagy (CMA) in which a cytosolic chaperone directs cargo to be translocated to the interior of lysosomes (Boya *et al.*, 2013). For the purposes of this chapter, autophagy will refer to macroautophagy, the non-selective turnover of cytosolic contents trapped in double-membraned vesicles.

Although autophagy had been observed by microscopy and studied biochemically for several years, a lack of genetic and molecular tools hampered studies to understand the regulation and control of this degradative process (Ohsumi, 2014). Genetic studies, pioneered by the Ohsumi and Klionsky groups in the 1990s, identified a large group of genes responsible for regulating autophagy in *Saccharomyces cerevisiae* (Harding *et al.*, 1995; Tsukada and Ohsumi, 1993). For many of these genes, their primary sequence did not reveal their function. However, one of these genes, *ATG8*, proved to encode a marker of autophagosomes which associates with nascent autophagosome membranes through a regulated lipidation event, and is essential for autophagosome formation (Ichimura *et al.*, 2004; Tanida, 2010). The identification of a specific membrane-associated protein from autophagosomes allowed researchers to mark the vesicles in live cells, and observe them without the need for electron microscopy (Klionsky *et al.*, 2016). The mammalian homologue of Atg8p, cellular MAP-LC3-B, was originally identified as a microtubule-associated protein and is commonly referred to as LC3-B or simply LC3 (Schaaf *et al.*, 2016) (Box 7.1). The phosphatidylethanolamine-conjugated form of LC3, known as LC3-II, is essential for the formation of autophagosomes and is associated with all autophagy-related membrane structures.

Box 7.1 Autophagy assays

The key issue is to distinguish between assays which analyse *autophagy*, which is defined as active degradation in autophagic vesicles, and assays which identify components of the autophagic protein machinery and their associations and modifications. This brief primer is designed to give an overview of the assays, but investigators are directed to the regularly updated article 'Guidelines for the use and interpretation of assays for monitoring autophagy' the latest version of which is referenced here (Klionsky *et al.*, 2016).

The gold standard for confirmation of autophagic activity remains the direct electron microscopy observation of autophagic vesicles, defined as vesicles of 300–600 nM with two lipid bilayers and light-staining cytosolic contents (Tanida, 2011). Double-membraned vesicles can be difficult to observe, as the distance between the two bilayers is small, but use of high-pressure freezing techniques or high-contrast staining techniques allow for reliable observation of double-membraned vesicles (Dales *et al.*, 1965; Suhy *et al.*, 2000). The ratio of light-staining autophagosomes and darker-staining autolysosomes has been used as a measure of progress through the pathway as well (Eskelinen, 2005).

Many of the most common assays involve the seven members of the cellular LC3 protein family. There are four LC3 isoforms, two of LC3A, and one each of LC3B and LC3C, encoded by the ATG8A,B, and C genes (Schaaf *et al.*, 2016). In addition, there are the family members GABARAP1 and GABARAPL1 and L2. All family members are associated with the phagophore and are highly conserved among mammals. Most assays are performed with antibodies against LC3B, which, for brevity, we will simply refer to as LC3. LC3 is largely cytosolic in cells with low baseline autophagy. When autophagy is induced, a ubiquitin-like conjugation system encoded by the ATG5-ATG12 proteins links LC3 to phosphatidylethanolamine, a conjugation which directs association of the protein with nascent 'omegasomes,' or autophagic membranes (Tanida, 2010).

Observation of LC3 by immunofluorescence microscopy usually requires transfected or integrated expression of a GFP–LC3 fusion, as few investigators have reported success using commercially available antibodies. Fusion expression plasmids are readily available from Addgene and commercial repositories. As autophagy signals increase, LC3 re-localizes from the cytosol to a punctate pattern. Quantification of these puncta, expressed either as number of cells displaying puncta or as number of puncta per cell, depending on the levels of background in the cells used, is one of the simplest assays for an increase in autophagosomes (Klionsky *et al.*, 2016). In addition, LC3 modification can be directly observed on a Western blot, since the additional charge of a phosphatidylethanolamine modification causes the 18kDa LC3 to run at an apparent 16 kDa on standard SDS-PAGE. Comparing the levels of the unmodified form, 'LC3-I,' to the levels of lipid-modified 'LC3-II,' when normalized to a loading control, can provide information about autophagy signalling. However, neither LC3-II formation nor puncta counting are a reliable measure of active autophagy. This is because puncta or LC3-II levels can increase by two methods: modification of LC3-I into LC3-II or inhibition of normal degradation/recycling of LC3-II back into LC3-I. One option is to inhibit degradation by treating with a spectrum of lysosome inhibitors, such as a combination of E64 and pepstatin D, which inhibits LC3-II degradation and allows measurement of LC3-II generation for the duration of the treatment (Tanida *et al.*, 2005).

Since autophagosomes mature into acidic amphisomes, sensors of pH are often used to monitor autophagic flux. Commercial sensors, such as lysotracker, likely stain amphisomes as well as autolysosomes, but should be used with caution, as they are not specific to autophagic compartments (Chikte *et al.*, 2014; Niemann *et al.*, 2000). More useful are tandem GFP- and RFP-fused versions of LC3. GFP fluorescence is much more pH sensitive than RFP fluorescence (Mizushima *et al.*, 2010). Therefore, GFP fluoresces in relatively neutral autophagosomes, while RFP fluoresces in autophagosomes, acidic amphisomes, and lysosomes. Green–red colocalization is read as autophagosomes while red signal alone indicates amphisomes and lysosomes. In addition, some commercial kits have come onto the market purporting to measure autophagy. While they may have some utility, when it comes to kits with proprietary components, *caveat emptor.* In some cases, for example, the kits include degradation inhibitors, and therefore measure input into the pathway, and not flux (Cheong *et al.*, 2011; Chinnadurai *et al.*, 2015).

To more directly observe active autophagy, investigators turn to the Sequestosome 1 (SQSTM1) protein, which was originally and popularly known as p62. SQSTM1/p62 is an adapter protein which binds to and directs cargo to the autophagosome for degradation. In this process, SQSTM1 is degraded, and in situations in which autophagy is highly active, degradation often outpaces new production, making steady-state levels of SQSTM1 a reliable indicator of levels of autophagic degradation. Other autophagic cargo adaptors, such as NBR1, can also be used to monitor degradation, but SQSTM1 is by far the most widely used (Johansen and Lamark, 2011). LC3 modification in the absence of SQSTM1 degradation can indicate the induction of autophagosomes that fail to mature into autolysosomes, as has been observed for coxsackievirus (Kemball *et al*., 2010; Shi *et al*., 2015). However, for some viruses SQSTM1 and other cargo adapters are cleaved by viral proteases, meaning that steady-state levels of the full-length proteins do not measure degradative autophagy during infection (Shi *et al*., 2013, 2014).

In summary, direct observation of autophagosomes by EM, counting of LC3 puncta, Western blots of LC3 modification, and quantification of steady state levels of SQSTM1 can be combined to give a complete picture of autophagic signalling and activity in an infected cell.

Activation of the autophagic pathway, as shown in Fig. 7.1, can be identified by the increased presence of nascent crescent-shaped (in two dimensional images) or cup-shaped (in three dimensions) bodies known as pre-autophagic structures (PAS) in yeast, or omegasomes in most other organisms (Simonsen and Stenmark, 2008). These crescents expand and self-fuse, resulting in a double-membraned autophagosome, which captures cytosolic contents. Autophagosomes, which contain two distinct lipid bilayers and are decorated with LC3-II, are the most distinct physical hallmark of the autophagy pathway. Autophagosomes mature into amphisomes through fusion with endosomes, which deliver vacuolar ATPases and promote acidification. Acidic amphisomes then fuse with lysosomes to become autolysosomes, which are electron-dense and contain active lysosomal contents (Reggiori and Ungermann, 2017). Each step in development of the autophagosome is regulated; the presence of autophagosomes with cytosolic contents does not guarantee that those contents are being degraded by delivery to autolysosomes, and assays for autophagy are distinct from assays for the presence of autophagic signalling (see Box 7.1).

The relationship between *Enteroviruses* and autophagy

The Kirkegaard group first recognized that there is a relationship between autophagy and poliovirus. Extensive density centrifugation experiments revealed that vesicles induced by infection or by expression of the poliovirus proteins 2BC and 3A contain cellular markers from the endoplasmic reticulum and other organelles, which is consistent with an autophagic origin (Suhy *et al*., 2000). However, during infection the apparent circumference and buoyant density of the 2BC- and 3A- associated vesicles was distinct from that of ER markers. At the time, autophagosome-specific markers were not known. The conclusion was that PV-induced vesicles are autophagosome-like, and primarily derived from the ER.

Several studies, including many using DNA viruses, concluded that autophagy acts as an

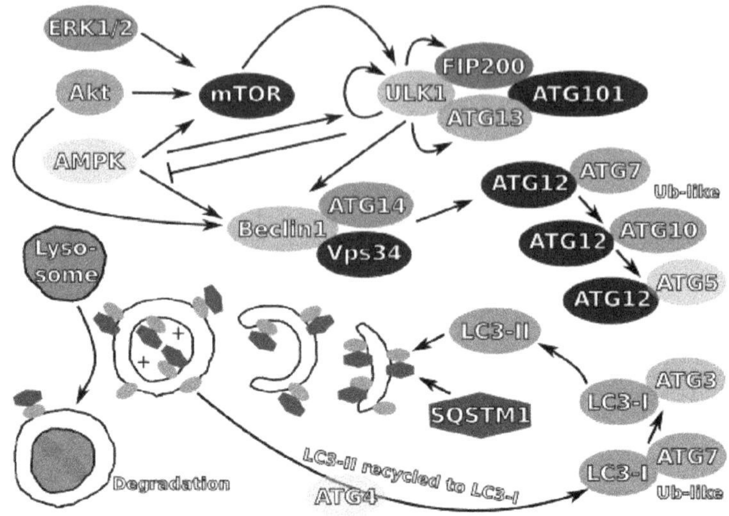

Figure 7.1 The Autophagic Pathway. A broad outline of some of the essential steps to generating an autophagosome is diagrammed here. Proteins discussed in the text of the chapter are emphasized; the reader is directed to the referenced papers for more detailed pathway analysis. Autophagy is regulated by signals being sent from ERK1/2, Akt, and AMPK to regulate autophagy levels based on nutrient levels, developmental signals, or other cellular stresses including infection (Hahn-Windgassen *et al.*; Kim *et al.*, 2011, 2014). Akt and AMPK can signal through mTOR or directly to Beclin 1 (Shang and Wang, 2011; Wang *et al.*, 2012; Zhang *et al.*, 2016). AMPK can directly phosphorylate ULK1, and there is a negative feedback phosphorylation event from ULK1 to AMPK (Kim *et al.*, 2011; Löffler *et al.*, 2011). mTOR signals to the ULK1 kinase, part of a multicomponent complex specifically regulating autophagy. ULK phosphorylates itself and other members of the complex on multiple sites, transmitting a signal to the Beclin1–Vps34–ATG14 complex (Alers *et al.*, 2012; Hara *et al.*, 2008; Hosokawa *et al.*, 2009; Puente *et al.*, 2016; Wirth *et al.*, 2013). Vps34, a phosphatidylinositol-3-Kinase, triggers at least two ubiquitin conjugation-like systems. One, ATG7/10, conjugates ATG5 to ATG12 to promote LC3 lipidation, and ATG7/3, which prepares the hATG8/LC3 protein to be added to phosphatidylethanolamine (Kim *et al.*, 2013; Nakatogawa, 2013). This lipidated form of LC3, called LC3-II, localizes to membranes, can be read as punctate staining in cell fluorescence assays, and runs at faster mobility on SDS-PAGE (Klionsky *et al.*, 2016). The cargo adapter SQSTM1 (p62) binds LC3 and is degraded when the autophagosome matures, then fuses with the lysosome (Ichimura *et al.*, 2008). Some LC3 is degraded as well, but LC3 on the cytosolic face of the autophagosome/autolysosome is thought to be recycled as LC3-I through the activity of the ATG4 protease (Satoo *et al.*, 2009).

antiviral host defence (Orvedahl and Levine, 2009; Yordy *et al.*, 2013). However, in the case of poliovirus it was shown that increased autophagy increases viral yield, while inhibition of autophagy, either through pharmacological agents or siRNA knockdown of genes essential for autophagy such as the LC3-encoding gene ATG8, inhibits virus replication (Jackson *et al.*, 2005). Autophagy has subsequently been shown to promote the replication of virtually all of the *Enteroviruses* in which autophagosomes have been detected: CVB3, CVB4, HRV-2, FMDV, EMCV, and EV71 (Huang *et al.*, 2009; Klein and Jackson, 2011; O'Donnell *et al.*, 2011; Wong *et al.*, 2008; Yoon *et al.*, 2008; Zhang *et al.*, 2011).

The question, then, revolved around the role autophagosome-like structures play in promoting virus replication. Data suggested that pre-lytic release of virus was autophagy-dependent, a process termed 'Autophagic Exit Without Lysis,' or AWOL. However, it was unclear exactly how autophagy, known for cytosolic degradation, could play a role in exit of cytosolic contents from the cell (Jackson *et al.*, 2005). Virus-induced autophagosome-like structures do not traffic along microtubules, as normal autophagosomes do, but are immobilized as long as microtubules are intact (Taylor *et al.*, 2009). Inhibition of microtubules leads to increased non-lytic release of virus, indicating that autophagosomes 'freed' from microtubules promote autophagy-mediated secretion of virus. But there was no known connection between autophagy and secretion.

Subsequent studies in *Pichia pastoris* and *Dictyostelium discoideum* showed the presence of a non-canonical secretory pathway dependent on autophagy (Duran *et al.*, 2010; Manjithaya *et al.*, 2010). While this pathway appears to be a very minor percentage of total cellular secretion and was initially only detectable using sensitive bioassays, it is now understood that autophagosomes, exosomes, and multivesicular bodies all intersect as secretory vessels (Amaya *et al.*, 2015; Bobrie *et al.*, 2011).

It is known that the poliovirus requires membranes as a site for RNA replication. This led to the hypothesis was that the virus uses the cytoplasmic face of autophagosomes for virus RNA replication (Schlegel *et al.*, 1996). However, it was also clear that autophagy must play a role in virus exit from the cell. Assays for extracellular virus, performed in the absence of detectable cell lysis, indicated that increasing the signal to the autophagy pathway increased release of virus (Jackson *et al.*, 2005). One school of thought is that virus exit primarily takes place through cell lysis (Knipe and Howley, 2007). However, for years published data suggested the idea of virus-containing vesicles 'budding' from the cell surface to release virus (Dunnebacke *et al.*, 1969). There are also functional reports containing evidence of non-lytic release, including studies of poliovirus exit from Caco-2 and K562 cells (Lloyd and Bovee, 1993; Tucker *et al.*, 1993). A role for components of the autophagy pathway in viral spread and pathogenesis was subsequently demonstrated using live cell imaging and transgenic mouse models. Autophagy components, including the LC3 protein, specifically promote non-lytic spread of the virus in cultured cells (Bird *et al.*, 2014). Autophagy also increases viral pathogenesis in a transgenic mouse expressing the human poliovirus receptor gene, perhaps the result of more rapid spread and intramuscular titres. However, despite the evidence that autophagy promoted extracellular virus release, a mechanism for release was unknown, so a model invoking exclusively lytic release of free virions predominated.

Release of *Enteroviruses* without lysis

However, the mystery of how a non-enveloped virus could be spreading from cell to cell without interfering with the plasma membrane remained. Double-membraned autophagosomes, when fusing with the plasma membrane, would leave behind a single membrane coat for these canonically non-enveloped virions. Alternatively, perhaps the viruses survive the conversion of autophagosomes into autolysosomes, and subsequent loss of one membrane, before fusing with the plasma membrane and being released as naked virions. The mystery of non-lytic release was partially solved when studies of Hepatitis A Virus (HAV), a related picornavirus, revealed a surprising finding: a portion of HAV is released in membranous vesicles (Feng *et al.*, 2013). Hepatitis A Virus does not appear to use an autophagy-related pathway for release, however. Rather, the ESCRT pathway plays a role in release of enveloped 'eHAV'. Even more intriguing was the finding that eHAV is still neutralized by antibodies to the capsid proteins by an as-yet unknown mechanism. It has been proposed that eHAV may be important for viral exit from hepatocytes that have specialized secretory pathways (Feng and Lemon, 2014). These findings led to the idea that picornaviruses may not be purely non-enveloped viruses and an explanation for cellular exit without lysis.

Coxsaskievirus B3, which induces autophagic signalling and autophagosomes but no autophagic degradation, is also released in vesicles (Robinson *et al.*, 2014). In the case of CVB3, these vesicles are decorated with the LC3 protein, indicating that they are derived from the autophagy pathway. Poliovirus was similarly shown to exit the host cell in vesicles decorated with autophagy markers, and the vesicles also contain phosphotidylserine, leading to a model that the lipid content of the autophagosome-like vesicles promotes phagocytosis of PV, CVB3, and rhinovirus by some cell types (Chen *et al.*, 2015). EV71 has subsequently been shown to be released non-lytically in an autophagy-dependent manner as well, indicating that this is a mechanism common across *Enteroviruses* (Too *et al.*, 2016).

Why do these viruses use the autophagy pathway for release, as opposed to the more canonical ER-Golgi secretory pathway utilized by many enveloped viruses? One clue is that the acidification of autophagic vesicles promotes the capsid maturation of viruses (Richards and Jackson, 2012). This maturation, a cleavage of the VP0 capsid protein which is required for infectivity, only occurs in viral particles containing genomes, although the specific mechanism is unknown. Therefore, entry of virus particles into autophagosomes may be a checkpoint promoting infectivity. Particles released in phosphatidylserine-rich autophagosome-like vesicles are more likely to be mature and infectious (Chen *et al.*, 2015). This appears to be the case for CVB3, HRV, and PV; however, FMDV yield is unaffected by Bafilomycin A1 treatment, indicating that acidification of a compartment is not required for replication or maturation of this virus (Gladue *et al.*, 2012).

Enterovirus proteins promote autophagic signalling and degradation

The poliovirus replication protein 3A, along with 2C one of two known membrane-associated proteins encoded by the virus, colocalizes with LC3 during infection (Echeverri *et al.*, 1998; Jackson *et al.*, 2005). 3A plays multiple roles during infection, including a role in viral RNA replication, but is not part of the virion itself. It has been shown for all positive-strand RNA viruses studied to date that genomic RNA replication occurs in association with cellular membranes, so the presence of 3A on autophagosomes led to the hypothesis that the autophagosome-like vesicles act as the physical substrate for genomic RNA replication. However, autophagosomes appear after the peak of RNA replication, making it unlikely that it is the primary purpose of the membranes (Belov *et al.*, 2007; Richards *et al.*, 2014).

Expression of the poliovirus membrane-associated protein 2BC induces large, single-membraned vesicles and lipid modification of LC3, but not double-membraned autophagosomes. Expression of poliovirus 3A, a transmembrane protein, induces ER distension, consistent with its ability to block ER-Golgi traffic, but no indications of autophagic signalling. Expression of 2BC and 3A together induces both LC3 modification and visible autophagosomes with autophagosome markers (Jackson *et al.*, 2005; Suhy *et al.*, 2000; Taylor and Kirkegaard, 2007). Since LC3 modification is thought to be sufficient for autophagy induction, the question is, what does 3A add, or what cellular function does 3A replace, to alter 2BC-induced single-membraned vesicles to double-membraned vesicles? A 3A mutant that fails to block ER-Golgi traffic also fails to mediate immobilization on microtubules, indicating that these functions are genetically linked (Taylor *et al.*, 2009).

More likely, however, is that 3A (and its precursor 3AB) have curvature effects on intracellular membranes much like the effect LC3 has on cellular membranes (Fujita *et al.*, 2007; Lama and Carrasco, 1996; Landajuela *et al.*, 2016; Towner *et al.*, 1996). The transmembrane domain, which is part of 3A, is not inserted into membranes using traditional ER-membrane-associated translation. 3AB is inserted into the outer leaflet of lipid bilayers, and this insertion induces positive membrane curvature (Fujita *et al.*, 2007; Towner *et al.*, 1996). 3AB appears to be capable of inducing curvature and invagination of vesicles independent of infection (Wang *et al.*, 2013). 3AB converts liposomes into multilamellar structures, including some which strongly resemble double-membraned autophagosomes, *in vitro.* Protein transfection of 3AB into HeLa cells induces vesicular structures and cytoplasmic invaginations (Wang *et al.*, 2013). These data indicate that viruses may not need to engage the cellular machinery at all in order to generate multilamellar membranes, although the autophagy machinery is important for successful infection.

The FMDV 2C protein interacts directly with the major autophagy regulating protein Beclin1 (Gladue *et al.*, 2012). The data indicate that this interaction interferes with autophagosome maturation, although it is unclear if this is a common mechanism for *Enterovirus* regulation. It is interesting to note that inhibition of ER-Golgi traffic in FMDV is regulated by 2B and 2C, not 3A as it is in some other picornaviruses, so it is possible that FMDV-2C and PV-3A are functional analogues in autophagy as well (Moffat *et al.*, 2007). The 2C protein of encephalomyocarditis virus (EMCV), along with the RNA-dependent RNA polymerase 3D, induce autophagic signalling, indicating that autophagy induction functions can be shared among different proteins among the *Enterovirus* genus (Hou *et al.*, 2014).

Triggering of autophagy upon virus entry

FMDV triggers autophagic signals upon entry of viral particles into cells, and LC3 puncta formation is not dependent on PI3-kinases, indicating that LC3 is being lipidated in a novel fashion. Autophagosome formation by FMDV is not dependent on the entering viral particle being replication-competent, indicating that the capsid proteins can trigger autophagy signals (Berryman *et al.*, 2012). EV71 capsid proteins can induce autophagic signals when expressed independently, indicating that capsid triggering of autophagy signals may be a common feature of some Enteroviruses (Hu *et al.*, 2016). Entry itself may be dependent on autophagy; echovirus 7 entry to cells requires several components of the autophagy machinery, although entering virus particles do not localize to autophagosomes themselves (Kim and Bergelson, 2014). It is not clear from the literature if echovirus 7 induces autophagy, however. The protein PLA2G16 may act as a determinant of whether entering virus-containing endosomes are targeted for autophagic degradation or survival (Staring *et al.*, 2017). A similar pathway has been described in clearance of intracellular bacteria. The reciprocal relationship between viral entry and autophagy is an area in which more research is needed.

Coxsackievirus regulation of the autophagy pathway

Coxsackievirus B3 has been the subject of extensive research regarding autophagy during infection. The virus shown to both induce autophagic signalling and benefit from the presence of autophagosomes, including the previously described 'megaphagosomes' (Kemball *et al.*, 2010). However, in some studies CVB3 and CVB4, appear to specifically inhibit autophagic degradation, as evidenced by accumulation of p62/SQSTM1 during infection in both *in vivo* and *ex vivo* models of infection (Shi *et al.*, 2015; Yoon *et al.*, 2008, 2009). This makes sense for the virus in a way, inducing the vesicles for a pro-viral purpose without the degradation of cytoplasmic contents which could act as a component of the innate immune response.

Several other lines of evidence indicate that coxsackievirus interacts with the autophagy pathway in a non-canonical fashion. Although multiple groups have reported no change in SQSTM1 levels bring coxsackievirus infections, it has been suggested that SQSTM1 is specifically cleaved during infection by the viral 2A protease (Shi *et al.*, 2013). Since SQSTM1 is a cargo adapter protein for autophagy, this would indicate a novel mechanism for inhibition of selective autophagy without inhibiting formation of autophagosomes or degradation in general. In fact, one of the SQSTM1 cleavage products is a dominant negative peptide, indicating that the virus actively inhibits autophagic cargo uptake (Fung *et al.*, 2013; Shi *et al.*, 2013). At least one alternative cargo receptor, NBR1, is also cleaved by the 2A and 3C viral proteases, indicating that the virus may have evolved more than one way to inhibit cargo delivery into the autophagosome (Shi *et al.*, 2014).

Expression of mutant forms of LC3 from recombinant CVB3 viruses provided evidence that LC3 can be used by the virus for processes not related to classical autophagy, possibly as a mechanism to generate membranes for RNA replication (Alirezaei *et al.*, 2015). These data are reminiscent of studies describing a non-canonical role for LC3 in coronavirus and arterivirus replication (de Haan and Reggiori, 2008).

The bactericidal/permeability-increasing protein (BPI) fold-containing family B, member 3 (BPIFB3) is an autophagy inhibitor which restricts CVB3 replication, but seems to do so independent of several components of the cellular autophagy machinery, including Beclin 1, ATG14, and ATG7 (Delorme-Axford *et al.*, 2014). Interestingly, BPIFB3 does not seem to restrict poliovirus replication, furthering the idea that PV and CVB3, two closely related viruses, have very different relationships with autophagy. Finally, RIP3, a cellular kinase that promotes autophagic degradation and associates with SQSTM1, also promotes CVB3 replication, and RIP3 is cleaved by the 3C protease late in viral infection (Harris *et al.*, 2015). One of these proteolytic products associates with the full-length protein, an interaction which is thought to lead to cell death. It is clear that coxsackievirus interacts with the autophagy pathway in several ways, including several which do not fit with the standard autophagy paradigm.

Enterovirus 71

Enterovirus 71 (EV71), an agent of hand, foot and mouth disease and which can cause mortality in children under 5, induces autophagic signalling *in vitro* and *in vivo* using cell culture and mouse models of infection (Huang *et al.*, 2009; Lee *et al.*, 2014). Autophagosomes induced by the virus has been reported to be beneficial to virus production in these models as well. In the case of NSC34 neuronal cells, EV71 infection is non-lytic, and exit of the virus is dependent on autophagy, mimicking what has been demonstrated for poliovirus (Too *et al.*, 2016). Studies have demonstrated that *bona fide* degradative autophagy results from infection by EV71, and the resultant amphisome/lysosome fusion may down-regulate apoptosis during infection (Xi *et al.*, 2013). However, it is unclear if autophagy always promotes EV71 replication. Autophagy in MRC5 cells, induced by peptide-mediated inhibition of the FLICE-like inhibitory protein (FLIP) appears to be an anti-EV71 pathway (Won *et al.*, 2012). One mechanism reported for regulation of autophagy during EV71 infection is down-regulation of microRNA miR-30, suggesting a novel route for viral control of the pathway (Fu *et al.*, 2015). Expression of EV71 capsid protein VP1 is also sufficient to induce ER stress resulting in *bona fide* autophagy in DIC5 brainstem neurons and HEK293 cells (Hu *et al.*, 2016).

Rhinovirus

Relatively little work has been done on human rhinovirus (HRV) and autophagy. There are, at present, 167 identified serotypes of HRV: 80 of species A, 32 of species B, and 55 of species C (Royston and Tapparel, 2016). These numbers are in flux and the *Picornaviridae* Pages website, maintained at the Pirbright Institute, is an excellent resource for these and all picornaviruses (www.picornaviridae.com). It is known that at least some of these species trigger autophagic signalling. HRV-2 was thought not to induce autophagic signals, but a subsequent publication showed that HRV-2 does induce autophagic signals, and these signals benefit the virus (Brabec-Zaruba *et al.*, 2007; Klein and Jackson, 2011). HRV-14 and HRV-16 also induce autophagic signals (Jackson *et al.*, 2005). It is unclear to date if any of these viruses induce autophagic degradation. HRV-1A, however, does not appear to induce autophagic signals, apparently using a different mechanism of Golgi vesiculation to generate vesicles during infection (Quiner and Jackson, 2010).

The dual nature of autophagy

When studies of autophagy and virus replication were first being carried out, the question often being asked was: 'Is autophagy pro-viral or anti-viral?'(Kudchodkar and Levine, 2009). The answer to this question is a resounding yes. Almost certainly there are situations where autophagy acts as a component of innate immunity. Just as certainly there are conditions in which autophagy promotes virus replication and spread. The specific virus, cell type, and other conditions influence the role of autophagy in virus replication.

Nowhere is this more apparent than the discovery that infected trophoblasts along the placental barrier, when infected, trigger an anti-viral autophagy in neighbouring cells (Delorme-Axford *et al.*, 2013). This pro-autophagic state, mediated by exosomes containing microRNAs from the chromosome 19 microRNA cluster (C19MC), blocks replication of a wide variety of viruses, including many that, in other circumstances, have been shown to benefit from autophagy. The lone exception, cytomegalovirus (CMV), typically inhibits autophagy and may use C19MC to enhance its own replication. It is important to realize that context is everything, and that studies carried out in cell culture lines almost certainly fail to tell the entire story.

The current model

In summary, Enteroviruses induce the autophagic pathway in multiple ways, although they do not always induce *bona fide* autophagic degradation. The primary purpose, for the viruses, appears to be to promote virion maturation and exit into vesicle at the end of the infectious cycle. It is unclear what the purpose of these vesicles is, although they may play roles in immune avoidance, facilitating viral uptake in non-enteric cells, or promoting high MOI infections by delivering multiple virions at once. Almost two decades of molecular research on autophagy and Enteroviruses has led us to this model; the only certainty about the next two decades is that current ideas will be challenged, and new models will constantly emerge.

References

Alers, S., Löffler, A.S., Wesselborg, S., and Stork, B. (2012). The incredible ULKs. Cell Commun. Signal *10*, 7. https://doi.org/10.1186/1478-811X-10-7

Alirezaei, M., Flynn, C.T., Wood, M.R., Harkins, S., and Whitton, J.L. (2015). Coxsackievirus can exploit LC3 in both autophagy-dependent and -independent manners in vivo. Autophagy *11*, 1389–1407. https://doi.org/10.1080/15548627.2015.1063769

Amaya, C., Fader, C.M., and Colombo, M.I. (2015). Autophagy and proteins involved in vesicular trafficking. FEBS Lett. *589*, 3343–3353. https://doi.org/10.1016/j.febslet.2015.09.021

Belov, G.A., Altan-Bonnet, N., Kovtunovych, G., Jackson, C.L., Lippincott-Schwartz, J., and Ehrenfeld, E. (2007). Hijacking components of the cellular secretory pathway for replication of poliovirus RNA. J. Virol. *81*, 558–567.

Belov, G.A., Nair, V., Hansen, B.T., Hoyt, F.H., Fischer, E.R., and Ehrenfeld, E. (2012). Complex dynamic development of poliovirus membranous replication complexes. J. Virol. *86*, 302–312. https://doi.org/10.1128/JVI.05937-11

Berryman, S., Brooks, E., Burman, A., Hawes, P., Roberts, R., Netherton, C., Monaghan, P., Whelband, M., Cottam, E., Elazar, Z., et al. (2012). Foot-and-mouth disease virus induces autophagosomes during cell entry via a class III phosphatidylinositol 3-kinase-independent pathway. J. Virol. *86*, 12940–12953.

Bird, S.W., Maynard, N.D., Covert, M.W., and Kirkegaard, K. (2014). Nonlytic viral spread enhanced by autophagy components. Proc. Natl. Acad. Sci. U.S.A. *111*, 13081–13086. https://doi.org/10.1073/pnas.1401437111

Bobrie, A., Colombo, M., Raposo, G., and Théry, C. (2011). Exosome secretion: molecular mechanisms and roles in immune responses. Traffic *12*, 1659–1668. https://doi.org/10.1111/j.1600-0854.2011.01225.x

Boya, P., Reggiori, F., and Codogno, P. (2013). Emerging regulation and functions of autophagy. Nat. Cell Biol. *15*, 713–720. https://doi.org/10.1038/ncb2788

Brabec-Zaruba, M., Berka, U., Blaas, D., and Fuchs, R. (2007). Induction of autophagy does not affect human rhinovirus type 2 production. J. Virol. *81*, 10815–10817.

Chen, Y.H., Du, W., Hagemeijer, M.C., Takvorian, P.M., Pau, C., Cali, A., Brantner, C.A., Stempinski, E.S., Connelly, P.S., Ma, H.C., *et al.* (2015). Phosphatidylserine vesicles enable efficient en bloc transmission of enteroviruses. Cell *160*, 619–630. https://doi.org/10.1016/j.cell.2015.01.032

Cheong, H., Lindsten, T., Wu, J., Lu, C., and Thompson, C.B. (2011). Ammonia-induced autophagy is independent of ULK1/ULK2 kinases. Proc. Natl. Acad. Sci. U.S.A. *108*, 11121–11126. https://doi.org/10.1073/pnas.1107969108

Chikte, S., Panchal, N., and Warnes, G. (2014). Use of LysoTracker dyes: a flow cytometric study of autophagy. Cytometry A *85*, 169–178. https://doi.org/10.1002/cyto.a.22312

Chinnadurai, R., Copland, I.B., Ng, S., Garcia, M., Prasad, M., Arafat, D., Gibson, G., Kugathasan, S., and Galipeau, J. (2015). Mesenchymal Stromal Cells Derived From Crohn's Patients Deploy Indoleamine 2,3-dioxygenase-mediated Immune Suppression, Independent of Autophagy. Mol. Ther. *23*, 1248–1261.

Dales, S., Eggers, H.J., Tamm, I., and Palade, G.E. (1965). Electron microscopic study of the formation of poliovirus. Virology *26*, 379–389.

de Haan, C.A., and Reggiori, F. (2008). Are nidoviruses hijacking the autophagy machinery? Autophagy *4*, 276–279.

Delorme-Axford, E., Donker, R.B., Mouillet, J.-F., Chu, T., Bayer, A., Ouyang, Y., Wang, T., Stolz, D.B., Sarkar, S.N., Morelli, A.E., et al. (2013). Human placental trophoblasts confer viral resistance to recipient cells. Proc. Natl. Acad. Sci. U.S.A. *110*, 12048–12053.

Delorme-Axford, E., Morosky, S., Bomberger, J., Stolz, D.B., Jackson, W.T., and Coyne, C.B. (2014). BPIFB3 regulates autophagy and coxsackievirus B replication through a noncanonical pathway independent of the core initiation machinery. MBio 5, e02147.

Dunnebacke, T.H., Levinthal, J.D., and Williams, R.C. (1969). Entry and release of poliovirus as observed by electron microscopy of cultured cells. J. Virol. *4*, 505–513.

Duran, J.M., Anjard, C., Stefan, C., Loomis, W.F., and Malhotra, V. (2010). Unconventional secretion of Acb1 is mediated by autophagosomes. J. Cell Biol. *188*, 527–536. https://doi.org/10.1083/jcb.200911154

Echeverri, A., Banerjee, R., and Dasgupta, A. (1998). Amino-terminal region of poliovirus 2C protein is sufficient for membrane binding. Virus Res. *54*, 217–223.

Eskelinen, E.L. (2005). Maturation of autophagic vacuoles in Mammalian cells. Autophagy *1*, 1–10.

Feng, Z., and Lemon, S.M. (2014). Peek-a-boo: membrane hijacking and the pathogenesis of viral hepatitis. Trends Microbiol. *22*, 59–64. https://doi.org/10.1016/j.tim.2013.10.005

Feng, Z., Hensley, L., McKnight, K.L., Hu, F., Madden, V., Ping, L., Jeong, S.H., Walker, C., Lanford, R.E., and Lemon, S.M. (2013). A pathogenic picornavirus acquires an envelope by hijacking cellular membranes. Nature *496*, 367–371. https://doi.org/10.1038/nature12029

Fu, Y., Xu, W., Chen, D., Feng, C., Zhang, L., Wang, X., Lv, X., Zheng, N., Jin, Y., and Wu, Z. (2015). Enterovirus 71 induces autophagy by regulating has-miR-30a expression to promote viral replication. Antiviral Res. *124*, 43–53. https://doi.org/10.1016/j.antiviral.2015.09.016

Fujita, K., Krishnakumar, S.S., Franco, D., Paul, A.V., London, E., and Wimmer, E. (2007). Membrane topography of the hydrophobic anchor sequence of poliovirus 3A and 3AB proteins and the functional effect of 3A/3AB membrane association upon RNA replication. Biochemistry *46*, 5185–5199.

Fung, G., Ng, C.S., Zhang, J., Shi, J., Wong, J., Piesik, P., Han, L., Chu, F., Jagdeo, J., Jan, E., et al. (2013). Production of a dominant-negative fragment due to G3BP1 cleavage contributes to the disruption of mitochondria-associated protective stress granules during CVB3 infection. PLOS ONE *8*, e79546.

Gladue, D.P., O'Donnell, V., Baker-Branstetter, R., Holinka, L.G., Pacheco, J.M., Fernandez-Sainz, I., Lu, Z., Brocchi, E., Baxt, B., Piccone, M.E., et al. (2012). Foot-and-mouth disease virus nonstructural protein 2C interacts with Beclin1, modulating virus replication. J. Virol. *86*, 12080–12090.

González-Polo, R.A., Pizarro-Estrella, E., Yakhine-Diop, S.M.S., Rodríguez-Arribas, M., Gómez-Sánchez, R., Casado-Naranjo, I., Bravo-San Pedro, J.M., and Fuentes, J.M. (2016). The Basics of Autophagy. In Autophagy Networks in Inflammation, M.C. Maiuri, and D. De Stefano, eds. (Springer International Publishing), pp. 3–20.

Hahn-Windgassen, A., Nogueira, V., Chen, C.-C., Skeen, J.E., Sonenberg, N., and Hay, N. Akt activates mTOR by regulating cellular ATP level and AMPK activity.

Hara, T., Takamura, A., Kishi, C., Iemura, S., Natsume, T., Guan, J.L., and Mizushima, N. (2008). FIP200, a ULK-interacting protein, is required for autophagosome formation in mammalian cells. J. Cell Biol. *181*, 497–510. https://doi.org/10.1083/jcb.200712064

Harb, J.M., and Burch, G.E. (1975). Spherical aggregates of coxsackie B4 virus particles in mouse pancreas. Beitr Pathol *156*, 122–127.

Harding, T.M., Morano, K.A., Scott, S.V., and Klionsky, D.J. (1995). Isolation and characterization of yeast mutants in the cytoplasm to vacuole protein targeting pathway. J. Cell Biol. *131*, 591–602.

Harris, K.G., Morosky, S.A., Drummond, C.G., Patel, M., Kim, C., Stolz, D.B., Bergelson, J.M., Cherry, S., and Coyne, C.B. (2015). RIP3 Regulates Autophagy and Promotes Coxsackievirus B3 Infection of Intestinal Epithelial Cells. Cell Host Microbe *18*, 221–232. https://doi.org/10.1016/j.chom.2015.07.007

Harrison, A.K., Bauer, S.P., and Murphy, F.A. (1972). Viral pancreatitis: ultrastructural pathological effects of Coxsackievirus B3 infection in newborn mouse pancreas. Exp. Mol. Pathol. *17*, 206–219.

Horne, R.W., and Nagington, J. (1959). Electron Microscope Studies of the Development and Structure of Poliomyelitis Virus. J. Mol. Biol. *1*, 333–338.

Hosokawa, N., Sasaki, T., Iemura, S., Natsume, T., Hara, T., and Mizushima, N. (2009). Atg101, a novel mammalian autophagy protein interacting with Atg13. Autophagy *5*, 973–979.

Hou, L., Ge, X., Xin, L., Zhou, L., Guo, X., and Yang, H. (2014). Nonstructural proteins 2C and 3D are involved in autophagy as induced by the encephalomyocarditis virus. Virol. J. *11*, 156. https://doi.org/10.1186/1743-422X-11-156

Hu, D.-D., Mai, J.-N., He, L.-Y., Li, P.-Q., Chen, W.-X., Yan, J.-J., Zhu, W.-D., Deng, L., Wei, D., Liu, D.-H., et al. (2016). Glucocorticoids Prevent Enterovirus 71 Capsid Protein VP1 Induced Calreticulin Surface Exposure by Alleviating Neuronal ER Stress. Neurotox. Res.

Huang, S.C., Chang, C.L., Wang, P.S., Tsai, Y., and Liu, H.S. (2009). Enterovirus 71-induced autophagy detected in vitro and in vivo promotes viral replication. J. Med. Virol. *81*, 1241–1252. https://doi.org/10.1002/jmv.21502

Ichimura, Y., Imamura, Y., Emoto, K., Umeda, M., Noda, T., and Ohsumi, Y. (2004). In vivo and in vitro reconstitution of Atg8 conjugation essential for autophagy. J. Biol. Chem. *279*, 40584–40592. https://doi.org/10.1074/jbc.M405860200

Ichimura, Y., Kominami, E., Tanaka, K., and Komatsu, M. (2008). Selective turnover of p62/A170/SQSTM1 by autophagy. Autophagy *4*, 1063–1066.

Jackson, W.T., Giddings, T.H., Taylor, M.P., Mulinyawe, S., Rabinovitch, M., Kopito, R.R., and Kirkegaard, K. (2005). Subversion of cellular autophagosomal machinery by RNA viruses. PLOS Biol. *3*, e156.

Johansen, T., and Lamark, T. (2011). Selective autophagy mediated by autophagic adapter proteins. Autophagy *7*, 279–296.

Kemball, C.C., Alirezaei, M., Flynn, C.T., Wood, M.R., Harkins, S., Kiosses, W.B., and Whitton, J.L. (2010). Coxsackievirus infection induces autophagy-like vesicles and megaphagosomes in pancreatic acinar cells in vivo. J. Virol. *84*, 12110–12124. https://doi.org/10.1128/JVI.01417-10

Kim, C., and Bergelson, J.M. (2014). Echovirus 7 entry into polarized caco-2 intestinal epithelial cells involves core components of the autophagy machinery. J. Virol. *88*, 434–443. https://doi.org/10.1128/JVI.02706-13

Kim, J., Kim, Y.C., Fang, C., Russell, R.C., Kim, J.H., Fan, W., Liu, R., Zhong, Q., and Guan, K.L. (2013). Differential regulation of distinct Vps34 complexes by AMPK in nutrient stress and autophagy. Cell *152*, 290–303. https://doi.org/10.1016/j.cell.2012.12.016

Kim, J., Kundu, M., Viollet, B., and Guan, K.L. (2011). AMPK and mTOR regulate autophagy through direct phosphorylation of Ulk1. Nat. Cell Biol. *13*, 132–141. https://doi.org/10.1038/ncb2152

Kim, J.H., Hong, S.K., Wu, P.K., Richards, A.L., Jackson, W.T., and Park, J.I. (2014). Raf/MEK/ERK can regulate cellular levels of LC3B and SQSTM1/p62 at expression levels. Exp. Cell Res. *327*, 340–352. https://doi.org/10.1016/j.yexcr.2014.08.001

Klein, K.A., and Jackson, W.T. (2011). Human rhinovirus 2 induces the autophagic pathway and replicates more efficiently in autophagic cells. J. Virol. *85*, 9651–9654. https://doi.org/10.1128/JVI.00316-11

Klionsky, D.J., Abdelmohsen, K., Abe, A., Abedin, M.J., Abeliovich, H., Acevedo Arozena, A., Adachi, H., Adams, C.M., Adams, P.D., Adeli, K., *et al.* (2016). Guidelines for the use and interpretation of assays for monitoring autophagy (3rd edition). Autophagy *12*, 1–222. https://doi.org/10.1080/15548627.2015.1100356

Knipe, D.M., and Howley, P.M. (2007). Fields' Virology (Lippincott Williams & Wilkins).

Kudchodkar, S.B., and Levine, B. (2009). Viruses and autophagy. Rev. Med. Virol. *19*, 359–378. https://doi.org/10.1002/rmv.630

Lama, J., and Carrasco, L. (1996). Screening for membrane-permeabilizing mutants of the poliovirus protein 3AB. J. Gen. Virol. *77*, 2109–2119. https://doi.org/10.1099/0022-1317-77-9-2109

Landajuela, A., Hervás, J.H., Antón, Z., Montes, L.R., Gil, D., Valle, M., Rodriguez, J.F., Goñi, F.M., and Alonso, A. (2016). Lipid Geometry and Bilayer Curvature Modulate LC3/GABARAP-Mediated Model Autophagosomal Elongation. Biophys. J. *110*, 411–422.

Lee, Y.R., Wang, P.S., Wang, J.R., and Liu, H.S. (2014). Enterovirus 71-induced autophagy increases viral replication and pathogenesis in a suckling mouse model. J. Biomed. Sci. *21*, 80. https://doi.org/10.1186/s12929-014-0080-4

Lloyd, R.E., and Bovee, M. (1993). Persistent infection of human erythroblastoid cells by poliovirus. Virology *194*, 200–209.

Löffler, A.S., Alers, S., Dieterle, A.M., Keppeler, H., Franz-Wachtel, M., Kundu, M., Campbell, D.G., Wesselborg, S., Alessi, D.R., and Stork, B. (2011). Ulk1-mediated phosphorylation of AMPK constitutes a negative regulatory feedback loop. Autophagy *7*, 696–706.

Manjithaya, R., Anjard, C., Loomis, W.F., and Subramani, S. (2010). Unconventional secretion of *Pichia pastoris* Acb1 is dependent on GRASP protein, peroxisomal functions, and autophagosome formation. J. Cell Biol. *188*, 537–546.

Mizushima, N., Yoshimori, T., and Levine, B. (2010). Methods in mammalian autophagy research. Cell *140*, 313–326. https://doi.org/10.1016/j.cell.2010.01.028

Moffat, K., Knox, C., Howell, G., Clark, S.J., Yang, H., Belsham, G.J., Ryan, M., and Wileman, T. (2007). Inhibition of the secretory pathway by foot-and-mouth disease virus 2BC protein is reproduced by coexpression of 2B with 2C, and the site of inhibition is determined by the subcellular location of 2C. J. Virol. *81*, 1129–1139.

Nakatogawa, H. (2013). Two ubiquitin-like conjugation systems that mediate membrane formation during autophagy. Essays Biochem. *55*, 39–50. https://doi.org/10.1042/bse0550039

Niemann, A., Takatsuki, A., and Elsässer, H.P. (2000). The lysosomotropic agent monodansylcadaverine also acts as a solvent polarity probe. J. Histochem. Cytochem. *48*, 251–258. https://doi.org/10.1177/002215540004800210

O'Donnell, V., Pacheco, J.M., LaRocco, M., Burrage, T., Jackson, W., Rodriguez, L.L., Borca, M.V., and Baxt, B. (2011). Foot-and-mouth disease virus utilizes an autophagic pathway during viral replication. Virology *410*, 142–150.

Ohsumi, Y. (2014). Historical landmarks of autophagy research. Cell Res. *24*, 9–23. https://doi.org/10.1038/cr.2013.169

Orvedahl, A., and Levine, B. (2009). Eating the enemy within: autophagy in infectious diseases. Cell Death Differ. *16*, 57–69. https://doi.org/10.1038/cdd.2008.130

Puente, C., Hendrickson, R.C., and Jiang, X. (2016). Nutrient-regulated Phosphorylation of ATG13 Inhibits Starvation-induced Autophagy. J. Biol. Chem. *291*, 6026–6035. https://doi.org/10.1074/jbc.M115.689646

Quiner, C.A., and Jackson, W.T. (2010). Fragmentation of the Golgi apparatus provides replication membranes for human rhinovirus 1A. Virology *407*, 185–195. https://doi.org/10.1016/j.virol.2010.08.012

Reggiori, F., and Ungermann, C. (2017). Autophagosome maturation and fusion. J. Mol. Biol. *429*, 486–496.

Richards, A.L., and Jackson, W.T. (2012). Intracellular vesicle acidification promotes maturation of infectious poliovirus particles. PLOS Pathog. *8*, e1003046.

Richards, A.L., Soares-Martins, J.A., Riddell, G.T., and Jackson, W.T. (2014). Generation of unique poliovirus RNA replication organelles. MBio *5*, e00833–13. https://doi.org/10.1128/mBio.00833-13

Robinson, S.M., Tsueng, G., Sin, J., Mangale, V., Rahawi, S., McIntyre, L.L., Williams, W., Kha, N., Cruz, C., Hancock, B.M., *et al.* (2014). Coxsackievirus B exits the host cell in shed microvesicles displaying autophagosomal markers. PLOS Pathog. *10*, e1004045. https://doi.org/10.1371/journal.ppat.1004045

Royston, L., and Tapparel, C. (2016). Rhinoviruses and Respiratory Enteroviruses: Not as Simple as ABC. Viruses *8*, E16. https://doi.org/10.3390/v8010016

Satoo, K., Noda, N.N., Kumeta, H., Fujioka, Y., Mizushima, N., Ohsumi, Y., and Inagaki, F. (2009). The structure of Atg4B-LC3 complex reveals the mechanism of LC3 processing and delipidation during autophagy. EMBO J. *28*, 1341–1350. https://doi.org/10.1038/emboj.2009.80

Schaaf, M.B., Keulers, T.G., Vooijs, M.A., and Rouschop, K.M. (2016). LC3/GABARAP family proteins: autophagy-(un)related functions. FASEB J. *30*, 3961–3978.

Schlegel, A., Giddings, T.H., Ladinsky, M.S., and Kirkegaard, K. (1996). Cellular origin and ultrastructure of membranes induced during poliovirus infection. J. Virol. *70*, 6576–6588.

Shang, L., and Wang, X. (2011). AMPK and mTOR coordinate the regulation of Ulk1 and mammalian autophagy initiation. Autophagy *7*, 924–926.

Shi, J., Fung, G., Piesik, P., Zhang, J., and Luo, H. (2014). Dominant-negative function of the C-terminal fragments of NBR1 and SQSTM1 generated during enteroviral infection. Cell Death Differ. *21*, 1432–1441.

Shi, J., Wong, J., Piesik, P., Fung, G., Zhang, J., Jagdeo, J., Li, X., Jan, E., and Luo, H. (2013). Cleavage of sequestosome 1/p62 by an enteroviral protease results in disrupted selective autophagy and impaired NFKB signaling. Autophagy *9*, 1591–1603. https://doi.org/10.4161/auto.26059

Shi, Y., He, X., Zhu, G., Tu, H., Liu, Z., Li, W., Han, S., Yin, J., Peng, B., and Liu, W. (2015). Coxsackievirus A16 elicits incomplete autophagy involving the mTOR and ERK pathways. PLOS ONE *10*, e0122109. https://doi.org/10.1371/journal.pone.0122109

Simonsen, A., and Stenmark, H. (2008). Self-eating from an ER-associated cup. J. Cell Biol. *182*, 621–622. https://doi.org/10.1083/jcb.200807061

Staring, J., von Castelmur, E., Blomen, V.A., van den Hengel, L.G., Brockmann, M., Baggen, J., Thibaut, H.J., Nieuwenhuis, J., Janssen, H., van Kuppeveld, F.J., *et al.* (2017). PLA2G16 represents a switch between entry and clearance of Picornaviridae. Nature *541*, 412–416. https://doi.org/10.1038/nature21032

Suhy, D.A., Giddings, T.H., and Kirkegaard, K. (2000). Remodeling the endoplasmic reticulum by poliovirus infection and by individual viral proteins: an autophagy-like origin for virus-induced vesicles. J. Virol. *74*, 8953–8965.

Tanida, I. (2010). Autophagosome Formation and Molecular Mechanism of Autophagy. Antioxid. Redox Signal.

Tanida, I. (2011). Autophagy basics. Microbiol. Immunol. *55*, 1–11. https://doi.org/10.1111/j.1348-0421.2010.00271.x

Tanida, I., Minematsu-Ikeguchi, N., Ueno, T., and Kominami, E. (2005). Lysosomal turnover, but not a cellular level, of endogenous LC3 is a marker for autophagy. Autophagy *1*, 84–91.

Taylor, M.P., and Kirkegaard, K. (2007). Modification of cellular autophagy protein LC3 by poliovirus. J. Virol. *81*, 12543–12553.

Taylor, M.P., Burgon, T.B., Kirkegaard, K., and Jackson, W.T. (2009). Role of microtubules in extracellular release of poliovirus. J. Virol. *83*, 6599–6609. https://doi.org/10.1128/JVI.01819-08

Too, I.H., Yeo, H., Sessions, O.M., Yan, B., Libau, E.A., Howe, J.L., Lim, Z.Q., Suku-Maran, S., Ong, W.Y., Chua, K.B., *et al.* (2016). Enterovirus 71 infection of motor neuron-like NSC-34 cells undergoes a non-lytic exit pathway. Sci Rep *6*, 36983. https://doi.org/10.1038/srep36983

Towner, J.S., Ho, T.V., and Semler, B.L. (1996). Determinants of membrane association for poliovirus protein 3AB. J. Biol. Chem. *271*, 26810–26818.

Tsukada, M., and Ohsumi, Y. (1993). Isolation and characterization of autophagy-defective mutants of *Saccharomyces cerevisiae*. FEBS Lett. *333*, 169–174.

Tucker, S.P., Thornton, C.L., Wimmer, E., and Compans, R.W. (1993). Vectorial release of poliovirus from polarized human intestinal epithelial cells. J. Virol. *67*, 4274–4282.

Wang, J., Ptacek, J.B., Kirkegaard, K., and Bullitt, E. (2013). Double-membraned liposomes sculpted by poliovirus 3AB protein. J. Biol. Chem. *288*, 27287–27298. https://doi.org/10.1074/jbc.M113.498899

Wang, R.C., Wei, Y., An, Z., Zou, Z., Xiao, G., Bhagat, G., White, M., Reichelt, J., and Levine, B. (2012). Akt-mediated regulation of autophagy and tumorigenesis through Beclin 1 phosphorylation. Science *338*, 956–959. https://doi.org/10.1126/science.1225967

Wirth, M., Joachim, J., and Tooze, S.A. (2013). Autophagosome formation – the role of ULK1 and Beclin1-PI3KC3 complexes in setting the stage. Semin. Cancer Biol. *23*, 301–309. https://doi.org/10.1016/j.semcancer.2013.05.007

Won, M., Jun, E.J., Khim, M., Hong, S.H., Park, N.H., Kim, Y.K., and Lee, H. (2012). Antiviral protection against enterovirus 71 mediated by autophagy induction following FLICE-inhibitory protein inactivation. Virus Res. *169*, 316–320.

Wong, J., Zhang, J., Si, X., Gao, G., Mao, I., McManus, B.M., and Luo, H. (2008). Autophagosome supports coxsackievirus B3 replication in host cells. J. Virol. *82*, 9143–9153. https://doi.org/10.1128/JVI.00641-08

Xi, X., Zhang, X., Wang, B., Wang, T., Wang, J., Huang, H., Wang, J., Jin, Q., and Zhao, Z. (2013). The interplays between autophagy and apoptosis induced by enterovirus 71. PLOS ONE *8*, e56966. https://doi.org/10.1371/journal.pone.0056966

Yoon, S.Y., Ha, Y.E., Choi, J.E., Ahn, J., Lee, H., and Kim, D.H. (2009). Autophagy in coxsackievirus-infected neurons. Autophagy *5*, 388–389.

Yoon, S.Y., Ha, Y.E., Choi, J.E., Ahn, J., Lee, H., Kweon, H.S., Lee, J.Y., and Kim, D.H. (2008). Coxsackievirus B4 uses autophagy for replication after calpain activation in rat primary neurons. J. Virol. *82*, 11976–11978. https://doi.org/10.1128/JVI.01028-08

Yordy, B., Tal, M.C., Hayashi, K., Arojo, O., and Iwasaki, A. (2013). Autophagy and selective deployment of Atg proteins in antiviral defense. Int. Immunol. *25*, 1–10. https://doi.org/10.1093/intimm/dxs101

Zhang, D., Wang, W., Sun, X., Xu, D., Wang, C., Zhang, Q., Wang, H., Luo, W., Chen, Y., Chen, H., *et al.* (2016). AMPK regulates autophagy by phosphorylating BECN1 at threonine 388. Autophagy *12*, 1447–1459. https://doi.org/10.1080/15548627.2016.1185576

Zhang, Y., Li, Z., Ge, X., Guo, X., and Yang, H. (2011). Autophagy promotes the replication of encephalomyocarditis virus in host cells. Autophagy *7*, 613–628.

The Lipid Blueprints of Replicating Viral Genomes

8

Nihal Altan-Bonnet[1*], Marianita Santiana[1] and Olha Ilnytska[2]

[1]Laboratory of Host-Pathogen Dynamics, National Heart Lung and Blood Institute, National Institutes of Health, Bethesda, MD, USA.
[2]Center for Lipid Research, Rutgers, The State University of New Jersey, New Brunswick, NJ, USA.

*Correspondence: nihal.altan-bonnet@nih.gov

https://doi.org/10.21775/9781910190739.08

Abstract
Plus-strand RNA viruses are the largest group of RNA viruses containing many human, animal and plant pathogens including viruses that are the causative agents of polio, dengue, yellow fever, hepatitis C, common cold and hoof and mouth disease. These viruses dramatically remodel their host cells' pre-existing organelles to generate new organelles on which they subsequently assemble their enzymes and replicate. Remarkably the replication organelles from many different viruses end up converging on common lipid blueprints, enriched in specific lipids such as phosphatidylinositol 4-phosphate (PI4P), phosphatidylethanolamine (PE) and cholesterol that facilitate the assembly of macromolecular replication complexes. In addition these lipids impart on the replication organelles a high degree of membrane curvature, which also likely aids the formation and function of the replication complexes. The goal of this article is to examine the mechanisms by which different viruses generate these common lipid blueprints, how they can facilitate replication and how targeting these lipids may be a novel and panviral therapeutic strategy.

Introduction
Plus-strand RNA viruses include many important human pathogens such as poliovirus (PV), coxsackievirus B3 (CVB3), hepatitis C virus (HCV), yellow fever virus, Dengue virus, West Nile virus and human rhinovirus (HRV). These viruses all remodel their hosts intracellular membranes to generate unique organelles with a distinct lipid and protein composition that is optimal for genome replication and virus assembly (Chen, 2015; Belov, 2013; Altan-Bonnet and Balla, 2012; Hsu *et al.*, 2010; den Boon and Alhlquist, 2010; Miller and Krijnse-Locker, 2008).

Once the plus-strand viral RNA genome slips into the cytoplasm, the host ribosomal machinery rapidly translates it resulting in the production of polytopic viral proteins. These polytopic viral proteins are subsequently proteolytically cleaved into structural capsid and/or envelope proteins as well as non-structural replication proteins. While the total number of different viral proteins synthesized by a single plus-strand RNA genome is small relative to DNA viruses or other microorganisms, RNA viruses are versatile and manage to pack a lot of functionality within a handful of proteins. One way they achieve this is by having precursors encode enzymatic functions unique from that of the final products. Another is by hijacking host proteins that have a wide set of cellular effectors. The latter enables viruses to transform existing host organelles including the endoplasmic reticulum (ER), the Golgi apparatus, trans-Golgi network (TGN), endosomes and mitochondria into so-called 'replication organelles' that have unique lipid blueprints which serve to copy viral RNA (de Boon and Ahlquist, 2010; Chen *et al.*, 2015; Hsu *et al.*, 2010; Belov and Sztul, 2014).

Membranes facilitate replication

Plus-strand RNA viruses harness different subcellular organelles to generate novel replication organelles: flock house virus and Tomato bushy stunt virus (TBSV) exploit the mitochondrial outer membrane and peroxisomes, respectively (Xu and Nagy 2015); brome mosaic virus (Diaz *et al.*, 2015), HCV (Alvisi *et al.*, 2011), Dengue virus (Welsch *et al.*, 2009) and West Nile virus all differentiate the ER into a membranous web (Mackenzie *et al.*, 2007; Kaufusi *et al.*, 2014); Rubella virus hijacks multiple secretory compartments as well as lysosomes (Fontana *et al.*, 2010); and Enteroviruses like PV, coxsackievirus and rhinovirus transform the entire secretory pathway into highly vesicular-tubular organelles (Belov *et al.*, 2007; Hsu *et al.*, 2010; Belov *et al.*, 2012; Richards *et al.*, 2014). Nearly always the cytoplasmic membrane leaflets of the remodelled organelles are utilized as platforms on which viral replication enzymes assemble.

So what are the advantages of replicating on membranes? While this is not fully understood, binding to membranes and constraining assembly and replication reactions to two-dimensional surfaces may increase the efficiency of enzymatic reactions (McCloskey and Poo, 1986). In particular in the early stages of viral infections when viral proteins are low in abundance, losing one degree of diffusional freedom may considerably increase the probability of viral proteins finding each other. Secondly, associating with membranes may favour a particular viral protein orientation and/or conformation that facilitate enzymatic reactions (Lomize *et al.*, 2007; Ilnystka *et al.*, 2013). Thirdly, the lipids themselves may be co-factors and stimulate enzymatic activity (Pogany and Nagy, 2015). Fourthly membranes with high degrees of curvature may generate pockets of cytoplasm where viral proteins can be concentrated (Belov *et al.*, 2012; Romero-Brey *et al.*, 2012; den Boon and Alhlquist, 2010). Finally these same highly curved membrane pockets could shield the viral replication machinery from the host innate immunity defences (Jensen and Thomsen, 2012).

Convergence on a common lipid blueprint

There are significant changes in the host cell lipid metabolism as a consequence of biogenesis of new replication organelles including ramping up long chain fatty acid import (Nchoutmboube, 2013), increasing overall lipid biosynthesis (Perera, 2012) and modulating phosphatidylinositide and sterol metabolism (Hsu *et al.*, 2010; Ilnytska *et al.*, 2013; Roulin *et al.*, 2014; Albulescu *et al.*, 2015). Remarkably the replication organelles of many different viruses end up converging on common lipid blueprints that are critical to initiating and sustaining viral replication. For example replication organelles from PV-, coxsackievirus-, enterovirus 71-, human rhinovirus (HRV)-, echovirus-, Aichi virus- and HCV-infected cells all have in common membranes enriched in phosphatidylinositol 4-phosphate (PI4P) and cholesterol lipids (Fig. 8.1) (Hsu, 2010; Ilnytska, 2013; Reiss *et al.*, 2011; Berger *et al.*, 2011; Roulin, 2014; Albulescu, 2015; Saito *et al.*, 2015); whereas those from TBSV, carnation Italian ring spot virus, cucumber necrosis tombus virus and nodamura insect virus, in addition to cholesterol, all have in common phosphatidylethanolamine (PE) lipids (Sharma *et al.*, 2010; Xu and Nagy, 2015).

Advantages of enriching for PI4P lipids in replication organelles

PI4P lipids are enriched to high levels in the replication organelles of many different plus-strand RNA viruses and have been reported to be critical for viral replication: depleting the PI4P lipids from the organelles by inhibiting their production or by converting them back to phosphatidylinositides, significantly blocks viral RNA synthesis.

Phosphatidylinositide phosphate lipids, while not as abundant as other cellular lipids like PE or phosphatidylcholine (PC) play important roles in uninfected cells as part of membrane-based scaffolds on which signalling, membrane remodelling and cytoskeletal machinery assemble and function. They can be rapidly produced by phosphatidylinositol kinases phosphorylating the inositol ring of phosphatidylinositide (PI) at different positions

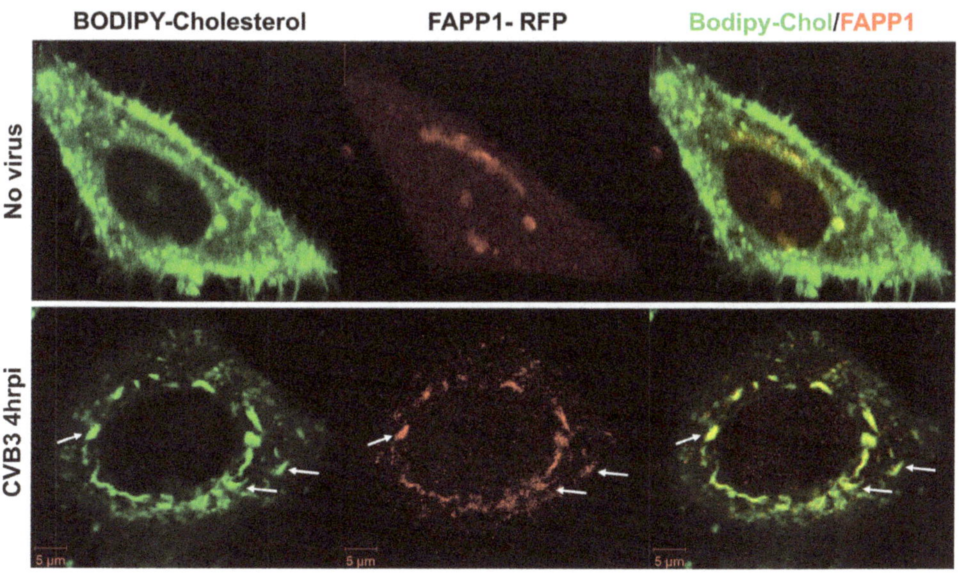

Figure 8.1 PI4P and cholesterol enriched replication organelle biogenesis in CVB3-infected cells. Cell is expressing FAPP1-mRFP (red), a fluorescently tagged reporter for PI4P lipids and is also labelled with BODIPY-cholesterol (green), a cholesterol mimic. The cell was infected with CVB3 for 4 hours and then imaged by confocal microscopy. Arrow points to replication organelles which are co-labelled with FAPP1-mRFP and BODIPY-cholesterol indicating the presence of both PI4P and cholesterol at these sites.

(Altan-Bonnet and Balla, 2012). PI4P lipids are normally produced at the plasma membrane and Golgi/TGN compartments of uninfected cells and recruit a wide repertoire of PI4P binding effectors (Tan and Brill, 2014) including four-phosphate adaptor proteins 1 and 2 (FAPP1 and FAPP2) which shape membranes (Godi *et al.*, 2004; Lenoir *et al.*, 2010); GOLPH3 proteins that link membranes to the actin cytoskeleton (Dippold *et al.*, 2009); ceramide transfer protein (CERT) and oxysterol binding proteins (OSBP) that further modify the lipid composition of the membrane (Toth *et al.*, 2006; Mesmin *et al.*, 2013); and coat protein adaptors such as GGA1 (Sun and Drubin 2012). As a result of the action of one or more of these PI4P effectors, the membrane domains can sort and concentrate cargo; initiate signalling; and become motile.

Even though the viral replication complexes are membrane bound many of the proteins synthesized by plus-strand RNA viruses are cytoplasmic and do not contain transmembrane or lipid anchoring domains. Binding to PI4P lipids could in principle facilitate membrane anchoring in addition to the protein–protein interactions with the few *bona fide* viral transmembrane proteins. Indeed, the PV RNA-dependent RNA polymerases (RdRp) which localize to PI4P enriched membrane domains within infected cells, have been demonstrated to selectively recognize and bind to PI4P lipid head groups over all other lipids including closely related phosphatidylinositide phosphates such as PI5P and PI (4,5) P_2 (Hsu *et al.*, 2010). Similarly, TBSV RdRps were shown to bind PE (Pogany and Nagy 2015), a lipid highly enriched in the peroxisome derived TBSV replication organelles and more recently, vaccinia, a DNA virus which replicates in the cytoplasm surrounded by ER derived membranes, was found to encode a replication protein with a PI4P binding domain (Kolli *et al.*, 2015).

In addition to anchoring viral proteins, PI4P lipids may facilitate replication by changing membrane structure either directly or indirectly, the latter through viral and host effectors. In artificial membrane systems, PI4P lipids in the absence of proteins, even at concentrations similar to those found in the cellular membranes of uninfected cells, induce high positive curvature membrane domains and generate complex three-dimensional structures (Furse *et al.*, 2012). Once generated, the binding of PV RdRps and other PI4P binding proteins could stabilize these high positive curvature domains. Moreover PV RdRps have been reported to polymerize and generate two-dimensional lattices (Lyle *et al.*, 2002), which could further enhance the formation and stability of high curvature PI4P membrane domains similar to what is observed for clathrin lattices which bind to PI (4,5)P_2-enriched positive curvature plasma membrane domains during endocytosis. In addition, host FAPP2 proteins have been shown to wedge themselves into the PI4P enriched membrane bilayers and induce tabulation (Lenoir *et al.*, 2010) (Fig. 8.2). These high curvature membrane domains stabilized by viral and host proteins could generate protected pockets of the cytosol where viral proteins and RNA templates could be concentrated to carry out replication reactions efficiently and protected from the host innate immunity defences. Consistent with this, depleting either PI4P lipids or FAPP2 proteins alters the morphology of replication organelles and inhibits viral replication (Khan *et al.*, 2014; Reiss *et al.*, 2011; Berger *et al*, 2011).

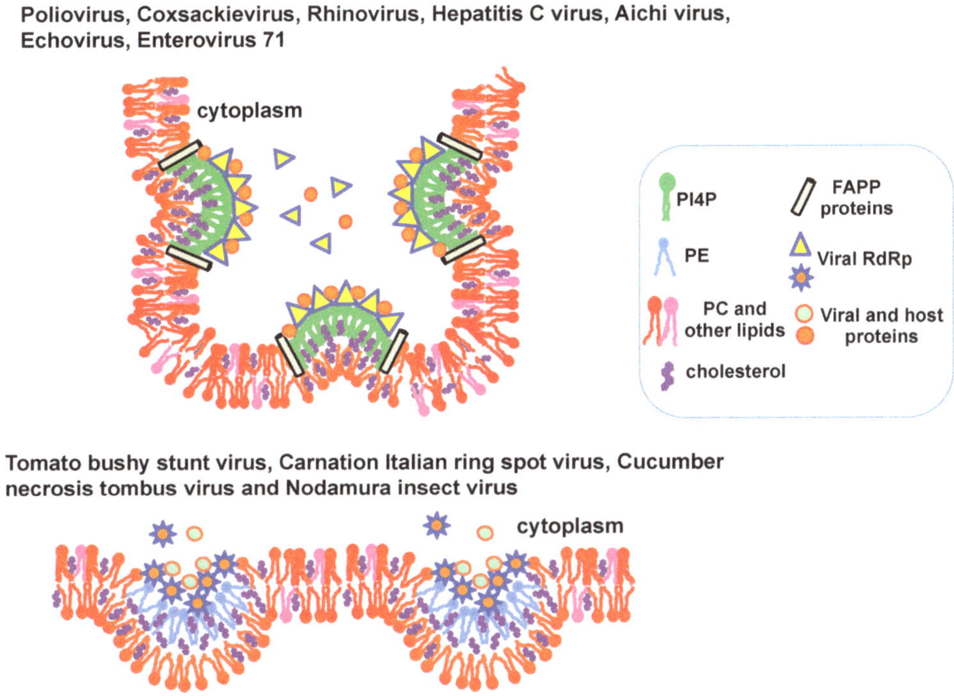

Figure 8.2 Schematic diagram of membrane domains enriched with PI4P or PE. High positive curvature PI4P/cholesterol domains and high negative curvature PE/cholesterol domains are exploited by a broad spectrum of positive strand RNA viruses. These PI4P/cholesterol and PE/cholesterol domains provide docking sites for viral and host proteins including the viral RdRps, as well generating pockets of cytoplasm where viral and host proteins can be concentrated and protected from the host's innate immunity defence components.

Mechanisms and consequences of viral induction of PI4P lipid production

PI4P lipids are produced by phosphorylation of PI by eukaryotic phosphatidylinositol 4 kinases (PI4K) (Balla and Balla, 2006; Altan-Bonnet and Balla, 2012; Clayton and Waugh, 2013). Mammalian cells contain four types of PI4Ks: the type II PI4 kinases alpha and beta, and the type III PI4 kinases alpha and beta (Balla and Balla, 2006). While all these enzymes produce the same product, PI4P, at steady state they localize to different subcellular organelles including plasma membrane, endosomes, Golgi/TGN and ER (Balla and Balla, 2006). All plus-strand RNA viruses that have been studied to date and that depend on PI4P lipids for replication harness the Type III PI4 kinases: for PV, CVB3, echovirus, enterovirus 71, HRVA2, HRVA16, HRVB14, HRVB37, and Aichi virus this is the phosphatidylinositol 4 kinase IIIβ (PI4KIIIβ); while for HRVA1A, HRVA2 and HCV strains it is mostly the PI4KIIIα enzyme (Hsu *et al.*, 2010; Reiss *et al.*, 2011; Arita *et al.*, 2011;

Greninger *et al.*, 2012; Altan-Bonnet and Balla, 2012; Ilnystka *et al.*, 2013; Roulin *et al.*, 2014; Ishikawa-Sasaki *et al.*, 2014).

In uninfected mammalian cells under steady-state conditions, PI4KIIIβ shuttles between the cytoplasm and the Golgi/TGN where it generates PI4P lipids. The PI4P lipids through recruitment of effectors regulate secretory cargo trafficking to the cell surface (Godi *et al.*, 2004). Apart from synthesizing PI4P, PI4KIIIβ itself has its own set of effectors, that bind to PI4KIIIβ independently of its kinase activity, including 14-3-3 proteins (Hausser *et al.*, 2006) and the Rab11 small GTPase, which regulates membrane trafficking among recycling endosomes, TGN and plasma membrane compartments (Burke *et al.*, 2014; Polevoy *et al.*, 2009).

Steady state PI4P pools at the Golgi/TGN provide an initial platform for PV and CVB3 replication (Hsu *et al.*, 2010) but as replication proceeds to exponential rates and viral protein levels rapidly rise, this Golgi/TGN-based platform becomes disassembled and replaced with new replication organelles emerging from ER exit sites (Hsu *et al.*, 2010). These new organelles are highly enriched in both PI4KIIIβ enzymes and PI4P lipids but notably do not contain COP1 or clathrin coat proteins (Hsu *et al.*, 2010). The lack of COPI coats is notable and has a profound effect on the fate of the Golgi as these proteins enable the membranes emerging out of the ER to recruit, retain, and traffic Golgi destined enzymes to the microtubule-organizing centre where they merge with the pre-exiting Golgi apparatus (Altan-Bonnet *et al.*, 2004). In the absence of this anterograde transport out of the ER, the ongoing retrograde transport of Golgi enzymes back to the ER leads to the eventual disassembly of the Golgi apparatus (Altan-Bonnet *et al.*, 2004). As a consequence of this, secretory trafficking of proteins including the major histocompatibility complex – which at the cell surface would be needed to alert other cells to the viral infection – becomes blocked (Deitz *et al.*, 2000; Choe *et al.*, 2005).

Both PI4KIIIβ and COP1 are effectors of the Arf1 GTPase (Godi *et al.*, 2004; Haynes *et al.*, 2007). The 3A proteins encoded by PV and coxsackievirus genomes, have lipophilic tails that anchor them to ER and Golgi/TGN membranes where they then interfere with COP1 recruitment by binding to Arf1's guanosine exchange factor GBF1 (Wessels, 2009; Hsu, 2010). Simultaneously 3A proteins also enhance PI4KIIIβ membrane recruitment (Hsu, 2010). While GBF1 can modulate the effector selectivity of Arf1 (Chen *et al.*, 2011), it is not known if enhancement of PI4KIIIβ recruitment is due to 3A binding to GBF1. Alternatively 3A proteins could be directly binding PI4KIIIβ (Dorobantu, 2015) or working through an intermediate host protein such as the acyl CoA-binding domain-containing 3 (ACBD3) protein which is reported to be part of the 3A–GBF1–PI4KIIIβ complex (Greninger, 2012; Sasaki *et al.*, 2012). Nevertheless, the result of 3A manipulation of COP1 and PI4KIIIβ recruitment is the emergence from the ER of PI4P enriched new organelles at a cost of disrupting secretory trafficking.

At peak replication times for CVB3 and PV infections, the total cellular PI4P levels are increased by ≈6-fold over those of uninfected cells (Hsu *et al.*, 2010) and nearly the entire intracellular pool of PI4P can be found at the replication organelles. The dramatic increase in PI4P levels is likely due to 3A proteins both enhancing PI4KIIIβ recruitment to membranes as well as stimulating its kinase activity (Ishikawa-Sasaki 2014). This is similar to what has been observed in HCV infections where viral NS5A proteins not only recruit PI4KIIIα to the ER but also stimulate its kinase activity (Riess *et al.*, 2011, Berger *et al.*, 2011). Unlike Enteroviruses, however, NS5A does not seem to disrupt COPI recruitment to any significant extent and notably the replication organelles of HCV are formed from domains of the ER *outside* the exit sites. The lack of interference with ER exit site activities leaves the Golgi apparatus and hence secretory trafficking largely intact in HCV infections. This is an important distinction between Enteroviruses and Hepaciviruses as the secretory pathway is vital to HCV export out of the cell whereas Enteroviruses exploit the autophagy pathway to non-lytically exit the cell (Jackson *et al.*, 2005; Chen *et al.*, 2015).

Phosphatidylethanolamine

Unlike PI4P, PE is one of the most abundant lipids in animal and plant cells, making up nearly 30% of total lipid mass. Through mechanisms yet unknown *Tombusviruses* TBSV and carnation Italian ring spot virus as well as the mosquito born nodamura virus all enrich for PE lipids at their replication sites by enhancing endogenous biosynthesis as well as redistributing pre-existing cellular PE pools (Xue and Nagy, 2015). Much like PI4P, PE appears to facilitate the assembly of TBSV proteins on peroxisome membranes and nodamura virus proteins on mitochondrial outer membranes (Xue and Nagy, 2015); binds and stimulates TBSV RdRp activity (Pogany and Nagy, 2015); and is a known inducer of negative membrane curvature (Farsad and de Camilli, 2003; Giang and Schick, 2014) and, thus, like PI4P, could facilitate replication by generating pockets of cytoplasm enclosed by membrane where replication machinery can be concentrated and protected. Why some plus-strand RNA viruses like TBSV and nodamura virus have evolved to depend on PE whereas others depend on PI4P is not understood and remains to be investigated.

Cholesterol: co-factor to stabilize PI4P and phosphatidylethanolamine domains

High levels of cholesterol lipids have been found within PI4P and PE enriched replication organelle membranes (Reghellin *et al.*, 2014; Ilnytska *et al.*, 2013; Sharma *et al.* 2010). Cholesterol from within the bilayer is an important regulator of membrane fluidity and the packing of proteins and other lipids (Jiang *et al.*, 2014). In particular cholesterol is known to be required for the formation and stability of any phosphatidylinositide phosphate domain larger than a few nanometres (Jiang *et al.*, 2014). Lipid domains solely composed of PI4P are not stable perhaps due to repulsive forces from their negatively charged head groups. Cholesterol promoting phosphatidylinositide phosphate clustering into large domains could enhance PI4P interactions with viral and host proteins similar to what has been reported for PTEN phosphatases whose binding to PI5P and PIP_2 lipid domains is significantly increased by the presence of cholesterol (Jiang *et al.*, 2014). In the case of PE, the cholesterol is also thought to help stabilize the induced negative curvature of membranes by ordering PE acyl chains (Giang and Schick, 2014). Supporting this idea that cholesterol may regulate viral replication by being a co-factor that increases PI4P and PE domain size and stabilizes membrane curvature, acutely depleting cholesterol from replication organelles in enteroviral infections, while PI4Ks are active and producing PI4P lipids, results in replication organelles that are smaller and fewer in number with altered morphology and significantly reduced replication (Ilnytska *et al.*, 2013). Similar findings have also been reported for HRV, HCV and TBSV replications (Ilnytska *et al.*, 2013; Roulin *et al.*, 2015). This impact of cholesterol on PI4P domain size and structure could also potentially explain why in its absence the rate of enteroviral 3CD protein processing is accelerated (Ilnytska *et al.*, 2013). 3CD in solution is unstable and quickly self-cleaves to generate 3C and 3D proteins (Marcotte *et al.*, 2007), the latter being the viral RdRp. However, some 3CD levels must be maintained throughout replication as 3CD's intrinsic primase and protease activities are required for priming RNA for elongation by 3D and modifying capsid proteins for packaging RNA (Marcotte *et al.*, 2007). In PV- and CVB3-infected cells, 3D and its precursor 3CD both bind PI4P/cholesterol enriched membrane domains and this likely helps attenuate the autocatalytic processing of 3CD (Ilnytska *et al.*, 2013). Consistent with this, in replication organelles with very high cholesterol levels, 3CD autocatalysis is highly attenuated (Ilnystka *et al.*, 2013). It is conceivable that binding to PI4P/cholesterol domains favours a particular 3CD conformation that makes its catalytic site either less active or less accessible to itself (for *in cis* proteolysis) or other 3CD proteins (for *in trans* proteolysis).

Viral mechanisms of obtaining cholesterol

Cholesterol homeostasis is a regulated complex process in which the cell, in response to its cholesterol needs, constantly modulates the endocytic uptake of cholesterol from the extracellular environment (bound to LDL or in micellar form); its biosynthesis at the ER; its trafficking to different subcellular compartments; and its storage in lipid droplets (Brown and Goldstein, 1986; Chang and Chang, 2008; Ikonen, 2008; Prinz, 2013).

At steady state, the plasma membrane is the largest reservoir for cholesterol in most mammalian cells, followed by endosomes and the Golgi/TGN compartments (Warnock *et al.*, 1993). In *Enterovirus*-infected cells there is a dramatic redistribution of plasma membrane cholesterol pools to the replication organelles emerging from ER exit sites (Ilnytska *et al.*, 2013). Live-cell time-lapse confocal imaging with fluorescent cholesterol mimics reveal that within 2 hours of infection the plasma membrane cholesterol pools are decreased by >90% with the cholesterol being internalized and redistributed to ER-derived replication organelles via Rab11-positive recycling endosomes (Ilnystka, 2013; Albulescu, 2015) (Fig. 8.3). Acutely depleting plasma membrane cholesterol pools and/or inhibiting endocytic accumulation of cholesterol either from the extracellular medium or from the plasma membrane, significantly inhibits PV and CVB3 replication along with disrupting organelle size and morphology (Ilnytska *et al.*, 2013; Roulin *et al.*, 2014, Albulescu *et al.*, 2015). Unlike many other cell types that rely on endocytic uptake of exogenous cholesterol, liver cells primarily rely on cholesterol biosynthesis to meet their needs and in HCV-infected liver cells the replication organelle cholesterol indeed appears to be largely derived from biosynthesis as when it is inhibited, replication organelle cholesterol levels are decreased and HCV replication is reduced (Saito *et al.*, 2015).

How does cholesterol traffic from recycling endosomes to the ER-derived replication organelles in *Enterovirus*-infected cells? One mechanism proposed involves complex formation between PI4KIIIβ and its effector Rab11 resulting in fusion of recycling endosomes with replication organelles (Ilnystka *et al.*, 2013). During infection with PV or CVB3, complex formation between Rab11 and PI4KIIIβ is enhanced >4-fold over uninfected cells even though levels of each protein are unchanged (Ilnystka *et al.*, 2013). Furthermore, ectopic expression of PV or CVB3 3A proteins alone is sufficient to enhance both PI4KIIIβ and Rab11 recruitment to membranes (Hsu *et al.*, 2010; Ilnytska *et al.*, 2013). However, a direct test of the role of the Rab11–PI4KIIIβ interaction in cholesterol delivery will involve

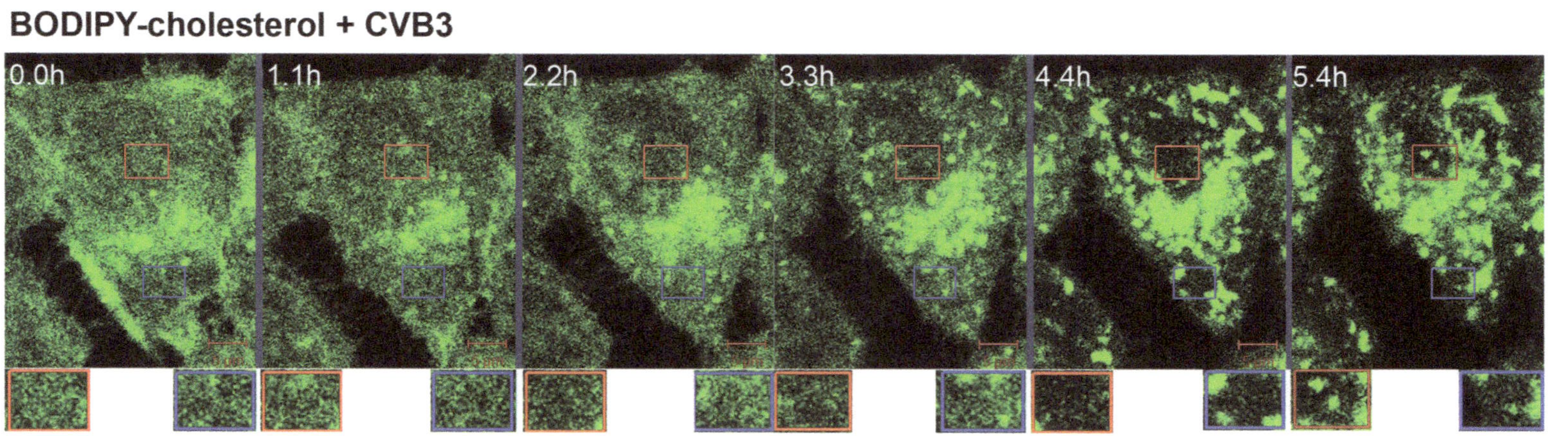

Figure 8.3 Redistribution of plasma membrane cholesterol pools to replication organelles in CVB3-infected cell. Cells were pulsed with BODIPY-cholesterol for 5 minutes and further incubated for 30 minutes to reach a steady state distribution prior to being infected with CVB3 and imaged every 5 minutes for 6 hours. Images show that the plasma membrane is being depleted of cholesterol while replication organelles are forming and enriching in cholesterol (red and blue boxes).

experiments mutating the specific amino acids at the interaction interface (Burke *et al.*, 2014). Alternatively, non-vesicular cholesterol transfer proteins such as the Osh and oxysterol binding proteins (OSBP) operating between plasma membrane, recycling endosomes and the ER and/or ER derived replication organelles may promote the transfer of cholesterol and this may be coupled to the presence of PI4P lipids (Mesmin *et al.*, 2013). In yeast cells, Osh proteins regulate cholesterol transfer to TBSV replication sites, where it stimulates replication (Barajas *et al.*, 2014) and when OSBP proteins, PI4P effectors that swap cholesterol and PI4P between membrane domains, are depleted, HCV and HRV replication is inhibited (Roulin *et al.*, 2014; Wang *et al.*, 2014). Consistent with this pharmacological inhibitors of PI4KIIIα recruitment by NS5A decrease not only the PI4P pools at replication organelles but also the cholesterol pools (Reghellin *et al.*, 2014).

Therapeutic potential of targeting lipids and future directions

In vitro studies inhibiting PI4KIIIβ or PI4KIIIα have shown promise as panviral therapeutic strategies with minimal cell toxicity (Hsu *et al.*, 2010, Reis *et al.*, 2011; Arita *et al.*, 2011; van der Schaar *et al.*, 2013; Ford-Siltz *et al.*, 2014; Chukkapalli *et al.*, 2015). This may be due to other PI4 kinase family members not virally hijacked meeting the hosts PI4P lipid needs, which relative to the viruses are significantly less. However, *in vivo* studies have revealed unexpected toxicities of interfering with PI4KIIIβ or PI4KIIIα kinase activities (LaMarche *et al.*, 2012; Spickler *et al.*, 2013) and it remains to be determined if ultimately these enzymes will be relevant as panviral drug targets. More investigations are clearly needed to determine if the observed *in vivo* toxicities are directly due to the inhibition of the kinase in question or off target effects of the inhibitor and whether there can be short windows of inhibitor administration where minimal toxicity and maximum inhibition of viral replication can be achieved. Alternatively, elucidating the mechanism of viral PI4K capture may reveal a therapeutic strategy, which would spare inhibiting the kinase activity while still preventing the viruses from hijacking the enzyme and inducing the production of large quantities of PI4P lipids.

Whether cholesterol manipulation can become a viable therapeutic strategy for controlling multiple viral infections also remains to be investigated. In addition to its role in the replication organelles, cholesterol is critical not only for the trafficking and dynamics of cellular receptors that many types of viruses rely on to gain entry into the cell but also in more unusual ways such as reported for West Nile virus, where cholesterol redistribution from the plasma membrane to the replication organelles inhibits apoptotic signalling pathways, thereby prolonging the life of the cell to support more viral replication (Mackenzie *et al.*, 2007). Since cholesterol is vital for numerous bodily functions, modulating its levels systemically is likely to have negative consequences. But cholesterol uptake could be acutely lowered at specific locations within the body critical for viral replication. For example the cells lining gut are an initial entry point and major replication site for many Enteroviruses. Perturbing cholesterol homeostasis in these cells by inhibiting endocytic uptake of cholesterol; or inhibiting Rab11 interaction with PI4KIIIβ; or blocking specific OSBP proteins operating in those cells are all possible antiviral therapeutic strategies. There is currently clinically approved therapeutics for patients with hypercholesterolaemia such as Ezetimibe, which block the endocytic uptake and transfer of dietary cholesterol to Rab11 recycling endosomes (Toth, 2010). It will be important to determine if Ezetimibe or similar therapeutics can lower the cholesterol levels in cells lining the gut sufficiently enough to inhibit enteroviral infection and propagation. In parallel careful epidemiological studies may reveal if there are correlations between patient cholesterol levels and propensity to be infected with cholesterol dependent Enteroviruses, HCV or West Nile virus.

We are often infected with multiple viruses at a given time and co-infections within single cells, especially between closely related RNA viruses, can give rise to genetic recombination, sharing of viral replication and structural proteins as well as increased circumvention of host innate immune defences (Debiaggi *et al.*, 2012). These coinfection phenomena are increasingly becoming an issue of concern for vaccine therapies and treatment strategies which are often targeted to be effective to only one specific virus (Debiaggi *et al.*, 2012; Holmblat *et al.*, 2014; Webster *et al.*, 2015). Identifying those lipid components that are critical to the replication of many different viruses and determining how they participate in replication can lead to panviral therapeutics, which have the potential to treat multiple viral infections at once. The strategy of targeting host proteins, such as those controlling lipid homeostasis, may help combat the emergence of drug resistance that inevitably arises from antivirals directly targeting viral enzymes. While viruses need large quantities of particular lipids for replication and have evolved to rely on specific proteins and pathways to procure those lipids, these same host proteins and pathways may prove to be optimal therapeutic targets as the host often has alternative redundant pathways from which to obtain the very same lipids to stay viable.

References

Albulescu, L., Wubbolts, R., van Kuppeveld, F.J., and Strating, J.R. (2015). Cholesterol shuttling is important for RNA replication of coxsackievirus B3 and encephalomyocarditis virus. Cell. Microbiol. *17*, 1144–1156. https://doi.org/10.1111/cmi.12425

Altan-Bonnet, N., and Balla, T. (2012). Phosphatidylinositol 4-kinases: hostages harnessed to build panviral replication platforms. Trends Biochem. Sci. *37*, 293–302. https://doi.org/10.1016/j.tibs.2012.03.004

Altan-Bonnet, N., Sougrat, R., and Lippincott-Schwartz, J. (2004). Molecular basis for Golgi maintenance and biogenesis. Curr. Opin. Cell Biol. *16*, 364–372. https://doi.org/10.1016/j.ceb.2004.06.011

Alvisi, G., Madan, V., and Bartenschlager, R. (2011). Hepatitis C virus and host cell lipids: an intimate connection. RNA Biol *8*, 258–269.

Arita, M., Kojima, H., Nagano, T., Okabe, T., Wakita, T., and Shimizu, H. (2011). Phosphatidylinositol 4-kinase III beta is a target of enviroxime-like compounds for antipoliovirus activity. J. Virol. *85*, 2364–2372. https://doi.org/10.1128/JVI.02249-10

Balla, A., and Balla, T. (2006). Phosphatidylinositol 4-kinases: old enzymes with emerging functions. Trends Cell Biol. *16*, 351–361.

Barajas, D., Xu, K., Sharma, M., Wu, C.Y., and Nagy, P.D. (2014). Tombusviruses upregulate phospholipid biosynthesis via interaction between p33 replication protein and yeast lipid sensor proteins during virus replication in yeast. Virology 471-473, 72–80. https://doi.org/10.1016/j.virol.2014.10.005

Belov, G.A., Altan-Bonnet, N., Kovtunovych, G., Jackson, C.L., Lippincott-Schwartz, J., and Ehrenfeld, E. (2007). Hijacking components of the cellular secretory pathway for replication of poliovirus RNA. J. Virol. *81*, 558–567.

Belov, G.A., Nair, V., Hansen, B.T., Hoyt, F.H., Fischer, E.R., and Ehrenfeld, E. (2012). Complex dynamic development of poliovirus membranous replication complexes. J. Virol. *86*, 302–312. https://doi.org/10.1128/JVI.05937-11

Belov, G.A., and Sztul, E. (2014). Rewiring of cellular membrane homeostasis by picornaviruses. J. Virol. *88*, 9478–9489. https://doi.org/10.1128/JVI.00922-14

Berger, K.L., Kelly, S.M., Jordan, T.X., Tartell, M.A., and Randall, G. (2011). Hepatitis C virus stimulates the phosphatidylinositol 4-kinase III alpha-dependent phosphatidylinositol 4-phosphate production that is essential for its replication. J. Virol. *85*, 8870–8883. https://doi.org/10.1128/JVI.00059-11

Brown, M.S., and Goldstein, J.L. (1986). A receptor-mediated pathway for cholesterol homeostasis. Science *232*, 34–47.

Burke, J.E., Inglis, A.J., Perisic, O., Masson, G.R., McLaughlin, S.H., Rutaganira, F., Shokat, K.M., and Williams, R.L. (2014). Structures of PI4KIIIβ complexes show simultaneous recruitment of Rab11 and its effectors. Science *344*, 1035–1038. https://doi.org/10.1126/science.1253397

Chang, T.Y., and Chang, C. (2008). Ezetimibe blocks internalization of the NPC1L1/cholesterol complex. Cell Metab. *7*, 469–471. https://doi.org/10.1016/j.cmet.2008.05.001

Chen, S., Cai, H., Park, S.K., Menon, S., Jackson, C.L., and Ferro-Novick, S. (2011). Trs65p, a subunit of the Ypt1p GEF TRAPPII, interacts with the Arf1p exchange factor Gea2p to facilitate COPI-mediated vesicle traffic. Mol. Biol. Cell *22*, 3634–3644. https://doi.org/10.1091/mbc.E11-03-0197

Chen, Y.H., Du, W., Hagemeijer, M.C., Takvorian, P.M., Pau, C., Cali, A., Brantner, C.A., Stempinski, E.S., Connelly, P.S., Ma, H.C., *et al.* (2015). Phosphatidylserine vesicles enable efficient en bloc transmission of enteroviruses. Cell *160*, 619–630. https://doi.org/10.1016/j.cell.2015.01.032

Choe, S.S., Dodd, D.A., and Kirkegaard, K. (2005). Inhibition of cellular protein secretion by picornaviral 3A proteins. Virology *337*, 18–29.

Chukkapalli, V., Berger, K.L., Kelly, S.M., Thomas, M., Deiters, A., and Randall, G. (2015). Daclatasvir inhibits hepatitis C virus NS5A motility and hyper-accumulation of phosphoinositides. Virology *476*, 168–179. https://doi.org/10.1016/j.virol.2014.12.018

Debiaggi, M., Canducci, F., Ceresola, E.R., and Clementi, M. (2012). The role of infections and coinfections with newly identified and emerging respiratory viruses in children. Virol. J. *9*, 247. https://doi.org/10.1186/1743-422X-9-247

Deitz, S.B., Dodd, D.A., Cooper, S., Parham, P., and Kirkegaard, K. (2000). MHC I-dependent antigen presentation is inhibited by poliovirus protein 3A. Proc. Natl. Acad. Sci. U.S.A. *97*, 13790–13795. https://doi.org/10.1073/pnas.250483097

den Boon, J.A., and Ahlquist, P. (2010). Organelle-like membrane compartmentalization of positive-strand RNA virus replication factories. Annu. Rev. Microbiol. *64*, 241–256. https://doi.org/10.1146/annurev.micro.112408.134012

Diaz, A., Zhang, J., Ollwerther, A., Wang, X., and Ahlquist, P. (2015). Host ESCRT proteins are required for bromovirus RNA replication compartment assembly and function. PLOS Pathog. *11*, e1004742. https://doi.org/10.1371/journal.ppat.1004742

Dippold, H.C., Ng, M.M., Farber-Katz, S.E., Lee, S.K., Kerr, M.L., Peterman, M.C., Sim, R., Wiharto, P.A., Galbraith, K.A., Madhavarapu, S., *et al.* (2009). GOLPH3 bridges phosphatidylinositol-4- phosphate and actomyosin to stretch and shape the Golgi to promote budding. Cell *139*, 337–351. https://doi.org/10.1016/j.cell.2009.07.052

Dorobantu, C.M., Ford-Siltz, L.A., Sittig, S.P., Lanke, K.H., Belov, G.A., van Kuppeveld, F.J., and van der Schaar, H.M. (2015). GBF1- and ACBD3-independent recruitment of PI4KIIIβ to replication sites by rhinovirus 3A proteins. J. Virol. *89*, 1913–1918. https://doi.org/10.1128/JVI.02830-14

Farsad, K., and De Camilli, P. (2003). Mechanisms of membrane deformation. Curr. Opin. Cell Biol. *15*, 372–381.

Fontana, J., López-Iglesias, C., Tzeng, W.P., Frey, T.K., Fernández, J.J., and Risco, C. (2010). Three-dimensional structure of Rubella virus factories. Virology *405*, 579–591. https://doi.org/10.1016/j.virol.2010.06.043

Ford, Siltz, L.A., Viktorova, E.G., Zhang, B., Kouiavskaia, D., Dragunsky, E., Chumakov, K., Isaacs, L., and Belov, G.A. (2014). New small-molecule inhibitors effectively blocking picornavirus replication. J. Virol. *88*, 11091–11107. https://doi.org/10.1128/JVI.01877-14

Ford, Siltz, L.A., Viktorova, E.G., Zhang, B., Kouiavskaia, D., Dragunsky, E., Chumakov, K., Isaacs, L., and Belov, G.A. (2014). New small-molecule inhibitors effectively blocking picornavirus replication. J. Virol. *88*, 11091–11107. https://doi.org/10.1128/JVI.01877-14

Furse, S., Brooks, N.J., Seddon, A.M., Woscholski, R., Templer, R.H., Tate, E.W., Gaffney, P.R.J., and Ces, O. (2012). Lipid membrane curcature induced by distearoyl phosphatidylinositol 4-phosphate. Soft matter *8*, 3090-3093.

Giang, H., and Schick, M. (2014). How cholesterol could be drawn to the cytoplasmic leaf of the plasma membrane by phosphatidylethanolamine. Biophys. J. *107*, 2337–2344. https://doi.org/10.1016/j.bpj.2014.10.012

Godi, A., Di Campli, A., Konstantakopoulos, A., Di Tullio, G., Alessi, D.R., Kular, G.S., Daniele, T., Marra, P., Lucocq, J.M., and De Matteis, M.A. (2004). FAPPs control Golgi-to-cell-surface membrane traffic by binding to ARF and PtdIns(4)P. Nat. Cell Biol. *6*, 393–404. https://doi.org/10.1038/ncb1119

Greninger, A.L., Knudsen, G.M., Betegon, M., Burlingame, A.L., and Derisi, J.L. (2012). The 3A protein from multiple picornaviruses utilizes the golgi adaptor protein ACBD3 to recruit PI4KIIIβ. J. Virol. *86*, 3605–3616. https://doi.org/10.1128/JVI.06778-11

Hausser, A., Link, G., Hoene, M., Russo, C., Selchow, O., and Pfizenmaier, K. (2006). Phospho-specific binding of 14-3-3 proteins to phosphatidylinositol 4-kinase III beta protects from dephosphorylation and stabilizes lipid kinase activity. J Cell Sci. *119*, 3613-21.

Haynes, L.P., Sherwood, M.W., Dolman, N.J., and Burgoyne, R.D. (2007). Specificity, promiscuity and localization of ARF protein interactions with NCS-1 and phosphatidylinositol-4 kinase-III beta. Traffic *8*, 1080–1092.

Holmblat, B., Jégouic, S., Muslin, C., Blondel, B., Joffret, M.L., and Delpeyroux, F. (2014). Nonhomologous recombination between defective poliovirus and coxsackievirus genomes suggests a new model of genetic plasticity for picornaviruses. MBio *5*, e01119–14. https://doi.org/10.1128/mBio.01119-14

Hsu, N.Y., Ilnytska, O., Belov, G., Santiana, M., Chen, Y.H., Takvorian, P.M., Pau, C., van der Schaar, H., Kaushik-Basu, N., Balla, T., *et al.* (2010). Viral reorganization of the secretory pathway generates distinct organelles for RNA replication. Cell *141*, 799–811. https://doi.org/10.1016/j.cell.2010.03.050

Ikonen, E., and Jansen, M. (2008). Cellular sterol trafficking and metabolism: spotlight on structure. Curr. Opin. Cell Biol. *20*, 371–377. https://doi.org/10.1016/j.ceb.2008.03.017

Ilnytska, O., Santiana, M., Hsu, N.Y., Du, W.L., Chen, Y.H., Viktorova, E.G., Belov, G., Brinker, A., Storch, J., Moore, C., *et al.* (2013). Enteroviruses harness the cellular endocytic machinery to remodel the host cell cholesterol landscape for effective viral replication. Cell Host Microbe *14*, 281–293. https://doi.org/10.1016/j.chom.2013.08.002

Ishikawa-Sasaki, K., Sasaki, J., and Taniguchi, K. (2014). A complex comprising phosphatidylinositol 4-kinase IIIβ, ACBD3, and Aichi virus proteins enhances phosphatidylinositol 4-phosphate synthesis and is critical for formation of the viral replication complex. J. Virol. *88*, 6586–6598. https://doi.org/10.1128/JVI.00208-14

Jackson, W.T., Giddings, T.H., Taylor, M.P., Mulinyawe, S., Rabinovitch, M., Kopito, R.R., and Kirkegaard, K. (2005). Subversion of cellular autophagosomal machinery by RNA viruses. PLOS Biol. 3, e156.

Jensen, S., and Thomsen, A.R. (2012). Sensing of RNA viruses: a review of innate immune receptors involved in recognizing RNA virus invasion. J. Virol. *86*, 2900–2910. https://doi.org/10.1128/JVI.05738-11

Jiang, Z., Redfern, R.E., Isler, Y., Ross, A.H., and Gericke, A. (2014). Cholesterol stabilizes fluid phosphoinositide domains. Chem. Phys. Lipids *182*, 52–61. https://doi.org/10.1016/j.chemphyslip.2014.02.003

Kaufusi, P.H., Kelley, J.F., Yanagihara, R., and Nerurkar, V.R. (2014). Induction of endoplasmic reticulum-derived replication-competent membrane structures by West Nile virus non-structural protein 4B. PLOS ONE 9, e84040. https://doi.org/10.1371/journal.pone.0084040

Khan, I., Katikaneni, D.S., Han, Q., Sanchez-Felipe, L., Hanada, K., Ambrose, R.L., Mackenzie, J.M., and Konan, K.V. (2014). Modulation of hepatitis C virus genome replication by glycosphingolipids and four-phosphate adaptor protein 2. J. Virol. *88*, 12276–12295. https://doi.org/10.1128/JVI.00970-14

Kolli, S., Meng, X., Wu, X., Shengjuler, D., Cameron, C.E., Xiang, Y., and Deng, J. (2015). Structure-function analysis of vaccinia virus H7 protein reveals a novel phosphoinositide binding fold essential for poxvirus replication. J. Virol. *89*, 2209–2219. https://doi.org/10.1128/JVI.03073-14

Lamarche, M.J., Borawski, J., Bose, A., Capacci-Daniel, C., Colvin, R., Dennehy, M., Ding, J., Dobler, M., Drumm, J., Gaither, L.A., *et al.* (2012). Anti-hepatitis C virus activity and toxicity of type III phosphatidylinositol-4-kinase beta inhibitors. Antimicrob. Agents Chemother. *56*, 5149–5156. https://doi.org/10.1128/AAC.00946-12

Lomize, A.L., Pogozheva, I.D., Lomize, M.A., and Mosberg, H.I. (2007). The role of hydrophobic interactions in positioning of peripheral proteins in membranes. BMC Struct. Biol. *7*, 44.

Lyle, J.M., Bullitt, E., Bienz, K., and Kirkegaard, K. (2002). Visualization and functional analysis of RNA-dependent RNA polymerase lattices. Science *296*, 2218–2222. https://doi.org/10.1126/science.1070585

Mackenzie, J.M., Khromykh, A.A., and Parton, R.G. (2007). Cholesterol manipulation by West Nile virus perturbs the cellular immune response. Cell Host Microbe *2*, 229–239.

Marcotte, L.L., Wass, A.B., Gohara, D.W., Pathak, H.B., Arnold, J.J., Filman, D.J., Cameron, C.E., and Hogle, J.M. (2007). Crystal structure of poliovirus 3CD protein: virally encoded protease and precursor to the RNA-dependent RNA polymerase. J. Virol. *81*, 3583–3596.

McCloskey, M.A., and Poo, M.M. (1986). Rates of membrane-associated reactions: reduction of dimensionality revisited. J. Cell Biol. *102*, 88–96.

Mesmin, B., Bigay, J., Moser von Filseck, J., Lacas-Gervais, S., Drin, G., and Antonny, B. (2013). A four-step cycle driven by PI(4)P hydrolysis directs sterol/PI(4)P exchange by the ER-Golgi tether OSBP. Cell *155*, 830–843. https://doi.org/10.1016/j.cell.2013.09.056

Miller, S., and Krijnse-Locker, J. (2008). Modification of intracellular membrane structures for virus replication. Nat. Rev. Microbiol. *6*, 363–374. https://doi.org/10.1038/nrmicro1890

Nchoutmboube, J.A., Viktorova, E.G., Scott, A.J., Ford, L.A., Pei, Z., Watkins, P.A., Ernst, R.K., and Belov, G.A. (2013). Increased long chain acyl-Coa synthetase activity and fatty acid import is linked to membrane synthesis for development of picornavirus replication organelles. PLOS Pathog. *9*, e1003401. https://doi.org/10.1371/journal.ppat.1003401

Pogany, J., and Nagy, P.D. (2015). Activation of Tomato Bushy Stunt Virus RNA-Dependent RNA Polymerase by Cellular Heat Shock Protein 70 Is Enhanced by Phospholipids In Vitro. J. Virol. *89*, 5714–5723. https://doi.org/10.1128/JVI.03711-14

Polevoy, G., Wei, H.C., Wong, R., Szentpetery, Z., Kim, Y.J., Goldbach, P., Steinbach, S.K., Balla, T., and Brill, J.A. (2009). Dual roles for the Drosophila PI 4-kinase four wheel drive in localizing Rab11 during cytokinesis. J. Cell Biol. *187*, 847–858. https://doi.org/10.1083/jcb.200908107

Prinz, W.A. (2013). A bridge to understanding lipid droplet growth. Dev. Cell *24*, 335–336. https://doi.org/10.1016/j.devcel.2013.02.004

Reghellin, V., Donnici, L., Fenu, S., Berno, V., Calabrese, V., Pagani, M., Abrignani, S., Peri, F., De Francesco R., and Neddermann, P. (2014). NS5A inhibitors impair NS5A-phosphatidylinositol 4-kinase IIIα complex formation and cause a decrease of phosphatidylinositol 4-phosphate and cholesterol levels in hepatitis C virus-associated membranes. Antimicrob Agents Chemother. *58*, 7128-40.

Reiss, S., Rebhan, I., Backes, P., Romero-Brey, I., Erfle, H., Matula, P., Kaderali, L., Poenisch, M., Blankenburg, H., Hiet, M.S., *et al.* (2011). Recruitment and activation of a lipid kinase by hepatitis C virus NS5A is essential for integrity of the membranous replication compartment. Cell Host Microbe *9*, 32–45. https://doi.org/10.1016/j.chom.2010.12.002

Richards, A.L., Soares-Martins, J.A., Riddell, G.T., and Jackson, W.T. (2014). Generation of unique poliovirus RNA replication organelles. MBio *5*, e00833–13. https://doi.org/10.1128/mBio.00833-13

Romero-Brey, I., Merz, A., Chiramel, A., Lee, J.Y., Chlanda, P., Haselman, U., Santarella-Mellwig, R., Habermann, A., Hoppe, S., Kallis, S., *et al.* (2012). Three-dimensional architecture and biogenesis of membrane structures associated with hepatitis C virus replication. PLOS Pathog. *8*, e1003056. https://doi.org/10.1371/journal.ppat.1003056

Roulin, P.S., Lötzerich, M., Torta, F., Tanner, L.B., van Kuppeveld, F.J., Wenk, M.R., and Greber, U.F. (2014). Rhinovirus uses a phosphatidylinositol 4-phosphate/cholesterol counter-current for the formation of replication compartments at the ER-Golgi interface. Cell Host Microbe *16*, 677–690. https://doi.org/10.1016/j.chom.2014.10.003

Saito, K., Shirasago, Y., Suzuki, T., Aizaki, H., Hanada, K., Wakita, T., Nishijima, M., and Fukasawa, M. (2015). Targeting cellular squalene synthase, an enzyme essential for cholesterol biosynthesis, is a potential antiviral strategy against hepatitis C virus. J. Virol. *89*, 2220–2232. https://doi.org/10.1128/JVI.03385-14

Sasaki, J., Ishikawa, K., Arita, M., and Taniguchi, K. (2012). ACBD3-mediated recruitment of PI4KB to picornavirus RNA replication sites. EMBO J. *31*, 754–766. https://doi.org/10.1038/emboj.2011.429

Sharma, M., Sasvari, Z., and Nagy, P.D. (2010). Inhibition of sterol biosynthesis reduces tombusvirus replication in yeast and plants. J. Virol. *84*, 2270–2281. https://doi.org/10.1128/JVI.02003-09

Spickler, C., Lippens, J., Laberge, M.K., Desmeules, S., Bellavance, É., Garneau, M., Guo, T., Hucke, O., Leyssen, P., Neyts, J., Vaillancourt, F.H., Décor, A., O'Meara J., Franti, M., and Gauthier, A. (2013). Phosphatidylinositol 4-kinase III beta is essential for replication of human rhinovirus and its inhibition causes a lethal phenotype in vivo. Antimicrob Agents Chemother. *57*, 3358-68.

Sun, Y., and Drubin, D.G. (2012). Pik1-ing clathrin adaptors. Nat. Cell Biol. *14*, 231–232. https://doi.org/10.1038/ncb2448

Tan, J., and Brill, J.A. (2014). Cinderella story: PI4P goes from precursor to key signaling molecule. Crit. Rev. Biochem. Mol. Biol. *49*, 33–58. https://doi.org/10.3109/10409238.2013.853024

Tóth, B., Balla, A., Ma, H., Knight, Z.A., Shokat, K.M., and Balla, T. (2006). Phosphatidylinositol 4-kinase IIIbeta regulates the transport of ceramide between the endoplasmic reticulum and Golgi. J. Biol. Chem. *281*, 36369–36377.

Toth, P.P. (2010). Drug treatment of hyperlipidaemia: a guide to the rational use of lipid-lowering drugs. Drugs *70*, 1363–1379. https://doi.org/10.2165/10898610-000000000-00000

van der Schaar, H.M., Leyssen, P., Thibaut, H.J., de Palma, A., van der Linden, L., Lanke, K.H., Lacroix, C., Verbeken, E., Conrath, K., Macleod, A.M., *et al.* (2013). A novel, broad-spectrum inhibitor of enterovirus replication that targets host cell factor phosphatidylinositol 4-kinase IIIβ. Antimicrob. Agents Chemother. *57*, 4971–4981. https://doi.org/10.1128/AAC.01175-13

Wang, H., Perry, J.W., Lauring, A.S., Neddermann, P., De Francesco R., and Tai, A.W. (2014). Oxysterol-binding protein is a phosphatidylinositol 4-kinase effector required for HCV replication membrane integrity and cholesterol trafficking. Gastroenterology. *146*, 1373–1385.

Warnock, D.E., Roberts, C., Lutz, M.S., Blackburn, W.A., Young, W.W., and Baenziger, J.U. (1993). Determination of plasma membrane lipid mass and composition in cultured Chinese hamster ovary cells using high gradient magnetic affinity chromatography. J. Biol. Chem. *268*, 10145–10153.

Webster, D.P., Klenerman, P., and Dusheiko, G.M. (2015). Hepatitis C. Lancet *385*, 1124–1135. https://doi.org/10.1016/S0140-6736(14)62401-6

Welsch, S., Miller, S., Romero-Brey, I., Merz, A., Bleck, C.K., Walther, P., Fuller, S.D., Antony, C., Krijnse-Locker, J., and Bartenschlager, R. (2009). Composition and three-dimensional architecture of the dengue virus replication and assembly sites. Cell Host Microbe *5*, 365–375. https://doi.org/10.1016/j.chom.2009.03.007

Welsch, S., Miller, S., Romero-Brey, I., Merz, A., Bleck, C.K., Walther, P., Fuller, S.D., Antony, C., Krijnse-Locker, J., and Bartenschlager, R. (2009). Composition and three-dimensional architecture of the dengue virus replication and assembly sites. Cell Host Microbe *5*, 365–375. https://doi.org/10.1016/j.chom.2009.03.007

Wessels, E., Duijsings, D., Niu, T.K., Neumann, S., Oorschot, V.M., de Lange, F., Lanke, K.H., Klumperman, J., Henke, A., Jackson, C.L., *et al.* (2006). A viral protein that blocks Arf1-mediated COP-I assembly by inhibiting the guanine nucleotide exchange factor GBF1. Dev. Cell *11*, 191–201.

Xu, K., and Nagy, P.D. (2015). RNA virus replication depends on enrichment of phosphatidylethanolamine at replication sites in subcellular membranes. Proc. Natl. Acad. Sci. U.S.A. *112*, E1782–91. https://doi.org/10.1073/pnas.1418971112

Index

2A protease 26, 64, 94, 96, 103–104
3′ non-coding region (NCR) 24
3A protein 134
3C protease 4, 26, 94, 96, 103–105
3CD protein 135–136
3-methyladenine (3-MA) 16
5′ non-coding region (NCR) 24, 64
5′-3′ exoribonuclease 1 see Xrn1
60S ribosomes 101
7-methylguanosine (7mg) cap 24

A

Abl tyrosine kinase 15
ACBD3 134
Actin cytoskeleton 15
Acyl CoA-binding domain-containing 3 see ACBD3
Adaptive mutations 2–3
Adenosine-uridine (AU)-rich element RNA binding factor 1 see AUF1
Adenovirus 61
ADP-ribosylation factor 1 see Arf1
ADP-ribosylation factor-binding protein GGA1 see GGA1
Aevado, Ashley 82
Ago2 95–96
Akt 118
Altered particle 11
AMP-activated protein kinase see AMPK
AMPK 118
Andino, Raul 5
Annexin II 10
Antibodies, neutralizing 8
Antivirals 2, 45, 41, 59, 2324, 69, 73, 139
A-particle 11
APOBEC3 95
Apolipoprotein B mrna editing enzyme, catalytic polypeptide-like 3G see APOBEC3
Apoptosis 2, 38, 104, 107, 123
Arf1 14, 134
Argonaute2 see Ago2
ATG10 118
ATG12 116, 118
ATG13 118
ATG14 122
ATG5 116, 118

ATG7 118, 122
ATG8 see LC3
AUF1 29–34, 101
AU-rich element RNA-binding protein 1 see AUF1
Autophagic Exit Without Lysis (AWOL) 119
Autophagosomes 16, 113–124
Autophagy 2, 100–101, 108, 113–124
Autophagy assays 115–117
Autophagy-related genes see ATG genes

B

Bactericidal/permeability-increasing protein (BPI) fold-containing family B, member 3 (BPIFB3) 122
Bafilomycin A1 120
Basic local alignment search tool see BLAST
Batschelet, Eduard 70
Beclin1 121
BIG1 29–32
BLAST 55
Boar enterovirus 81
Bordería, Antonio 80
Bottlenecks, population 78–79, 82
Brefeldin A-inhibited guanine nucleotide-exchange protein see BIG1

C

Caco-2 cells 15
Cadherin Related Family Member 3 see CDHR3
Cajal bodies 93
Camp response element-binding protein phosphorylated (CREB-P) 26–27
Cancer, viral therapy against 5
Canyon region of viral capsid 8, 62–63
Cap-binding protein see eIF4e
Caprin1 94, 97
Capsid structure and maturation 8, 120
CAR 9
CARD 36
Casein kinase II 98–100
Caspase recruitment domain see CARD
Caveolar endocytosis 13
Caveolin-1 13, 15, 16
C-C chemokine receptor type 4 see Ccr4
Ccr4 40, 94–95
Cdc42 13–14
CDHR3 52–54, 61
Cell division control protein 42 see Cdc42
Cell stress 93–106, 133–124
Ceramide transfer protein see CERT
CERT 132
Chaperone-mediated autophagy 115
Cholesterol 16, 129, 135–139
Circular resequencing (cirseq) 82
C-Jun N-terminal kinase see JNK
Clathrin-independent carriers/GPI-enriched early endosomal compartments see CLIC and GEECs
Clathrin-mediated endocytosis 13 132–134
CLIC 13
CLRs 36
CLUSTAL 55–56
Coat protein complex see COPI and COPII
Coxsackie and Adenovirus Receptor see CAR

Codon deoptimization 4
Collaborative Cross 5
Complementation 82
Consensus (or average) sequence 82
COPI 134
COPII 29–32
Coronavirus 61
Coxsackie A9 virus, population dynamics 69
Coxsackievirus A2, A4, A10 recombination 74
Coxsackievirus A9, mutation rate of 71–72
Coxsackievirus B3, high fidelity variants of 72
Coxsackievirus B3, relative robustness of 74–76
Coxsackievirus B3, RNA replication 36–37
Coxsackievirus, replication organelles 130–131
Cpg dinucleotide 4
Cricket paralysis virus and stress granules 102
C-Type lectin receptors see Clrs
Cytomegalovirus (CMV) and autophagy 123

D

DAF 9–10, 15–16
Dcp1a 40, 94,96, 103, 107
DDX6 see Rck/p54
Decapping MRNA 1A see Dcp1a
Defective ribosomal products see DRIPs
Delay Accelerating Factor see DAF
Dexd/H-box helicase family 36
DIC5 brainstem neurons 123
Dictyostelium discoideum 119
Digital organisms 82
Dnastar 56
Domingo, Esteban 70
Double-membraned vesicles 113–124
Double-stranded RNA (dsRNA)-activated protein kinase R see PKR
DRIPs 101
Dynamin 13, 15, 16

E

Echovirus, and cell entry 12, 16
Edc3/4 40, 103–104
eIF2α 95
eIF2α-kinases 95
eIF3 26, 94
eIF4a 94
eIF4e 26, 94–95
eIF4g 26, 39, 94, 95, 98
Eigen, Manfred 70
Endocytosis 12–13
Endosomal maturation 16
Endosomal pH 14
Enhancer of mRNA decapping 3/4 see Edc3/4
Enterovirus 109 (HEV109) 81
Enterovirus 71 (EV71) 1, 10, 24, 35–37, 71, 122–123
  high fidelity variants 72
  recombination 73–74
  mutation rate 71
  receptors 10

Enterovirus D-68 (EVD-68) 1, 24
  receptors 10
Epistasis 82
ER exit sites 136
ER stress 123
Eradication of viruses 1, 3–4
ERK1/2 118
Error catastrophe 82
Eukaryotic initiation factors see eIF2α and eIF3 and eIF4a and eIF4e and eIF4g
EVH1 domain, of Dcp1a 107
Evolvability 77, 83
Exosome 102
Ezetimibe 138–139

F

FAPP1 and FAPP2 132
Far-upstream element-binding protein 1/2 see FBP1/2
FBP1/2 35
FIP200 see RB1CC1
Fitness variations and landscapes 71, 74–78, 83
FLICE-like inhibitory protein see FLIP
FLIP 123
Flock house virus replication organelles 130
Flotillins 14
Foot-and-mouth disease virus (FMDV) 1, 51
  and autophagy 121
  population dynamics 70–73, 79–80
Founder effects 71
Four-phosphate adaptor proteins 1/2 see FAPP1 and FAPP2
Fragment of antibody (Fab) capsid mapping 63
Fyn Tyrosine-protein kinase 14

G

G3BP 39
G3BP1 and G3BP2 94–102
G3BP1/2 39–40, 94–107
GABARAP1/L1/L2 116
Gamma-aminobutyric acid receptor-associated protein-like see GABARAP1/L1/L2
GBF1 29–32, 134
GCN2 95
GEECs 13
Gene chips 52
General control non-depressible 2 kinase see GCN2
Genetic robustness 75–77
Genome overview 24–25
GGA1 132
Glycophosphatidylinositol anchored protein see GPI
Golgi phosphoprotein 3 see GOLPH3
Golgi-specific brefeldin A-resistance guanine nucleotide exchange factor 1 see GBF1
GOLPH3 132
GPI 14
GPI-Enriched Endocytic Compartments see GEECs
Group XVI phospholipase A2 see PLA2G16
GW182 protein 94–96, 103

H

Haem-regulated kinase see HRI
Hamming distance 83

HBMEC 15
HDAC6 94
Heat shock protein 27 (Hsp27) 97
HEK293 cells 123
Heparan sulfate 10
Hepatitis A Virus (HAV) 1, 51, 119
Hepatitis C Virus (HCV) 2, 97, 129, 133, 135
Heterogeneous nuclear ribonucleoprotein A1 see hnRNP A1
Heterogeneous nuclear ribonucleoprotein K see hnRNP K
Heterogeneous nuclear ribonucleoproteins see hnRNPs
High-throughput sequencing 81
Histone deacetylase 6 see HDAC6
hnRNP A1 35
Hnrnp E1 see poly(C)-binding protein 1
hnRNP K 35
hnRNPs 28
HRI 95
Human 1000 Genomes Project 5
Human brain microvascular endothelial cells see HBMEC
Human Immunodeficiency Virus (HIV), sequence dynamics of 78
Hypercholesterolaemia 138–139

I

I kappa-B alpha see IκBα
ICAM-1/ ICAM-5 9–10, 52, 54, 62
Influenza A virus and stress granules 106
Integrin, as virus receptor 9
Intercellular adhesion molecule see ICAM-1/ ICAM-5
Interferon (IFN) 25, 36, 104
Interferon regulatory factor 7 see IRF7
Interferon response 35
Interferon stimulated gene (ISG) 104
Interferon-alpha/beta (IFNα/β) 104, 106
Internal ribosome entry site see IRES
International Committee on the Taxonomy of Viruses (ICTV) 52–53
Intestinal entry 17
Intrinsically disordered domain (ID) 100
IRES 24–25, 53, 64
IRES trans-acting factors see ITAFS
IRF7 106
ITAFS 24, 28
Italian ring spot virus 135
I-TASSER 57–59
Iterative Threading assembly Refinement see I-TASSER
IκBα 105

J

JNK 104

K

K nuclear shuttling (KNS) import pathway 28
Kew, Olin 4
Klionsky, Daniel 115
Knowles, Nick 53

L

La autoantigen 28, 35
Lasergene (dnastar) 56

Lauring, Adam 75
LC3 protein 29–33, 115–124
LDLR 11, 52, 54
Leader protein 64
Lethal mutagenesis 83
LGP2 36, 106
Lipid droplets 136
Lipids 129–138
Low complexity amino acid diversity (LC) domain 100
Low density lipoprotein receptor see LDLR
Lsm14A 103
Lsm1-7 94, 96
Lupus erythematosus 28
Lymphocytic choriomeningitis virus (LMCV) 79

M

Macropinocytosis 13–14
Master sequence 83
MAVS 36–38, 106
Mcintyre, Chloe 53
MDA5 36–37, 106
MEGA 54, 56
Melanoma differentiation-associated protein 5 see MDA5
Membrane fission 13
Membrane–virion complexes (MVCs) 114
Mengo virus 51
Mengo virus and stress granules 106
MEX3A E3 ubiquitin ligase 105
MHC presentation 113
MHC-I -associated GRP78 10
Microtubule-associated protein, light chain 3 see MAP-LC3
Minor, Phil 3
Minority variants in population 80
Mir-30 123
Mitochondrial Antiviral Signaling Protein see MAVS
Mitophagy 115
MODELLER 57–59
MRC5 cells 123
mRNA processing body assembly factor 1-7 see Lsm1-7
mRNA processing body assembly factor see Lsm14A
mRNA-binding proteins (RPBs) 94
mTOR 118
Muller's ratchet 71
MUSCLE 55–56
Mutant spectrum 83
Mutation frequency 83
Mutation hot-spots 81
Mutation rate 83
Mutational fitness effect 83
Mutational robustness 75–77
Mutator/antimutator 83

N

NBR1 117, 122
Necrosis 2
Necrostatins 2
Negative selection 83
Neighbor of BRCA1 gene 1 see NBR1

Neighbour-joining tree of known RV genotypes 53
Neutralizing antibody 63
Neutralizing immunogenic sites (Nims) 60–64
Newcastle-Disease virus and stress granules 105
Next-generation sequencing (NGS) 81
NF-κB 36, 104–105
NLRs 36
Nodamura virus 135
NOD-like receptors see NLRs
Non-lytic release 119–120
Not1 94–95
Novella, Isabel 77
NS5A, of HCV 134
NSC34 neuronal cells 123
NTF2 domain of G3BP1 and G3BP2 99–100
Nuclear factor κB see NF-κB
Nuclear localization signal (NLS) 28
Nuclear pore complexes (NPCs) 28
Nuclear speckles 93
Nucleolin 10
Nucleoporins see Nup62/153/98
Nup62/153/98 28

O

OAS 106
Ohsumi, Yoshinori 115
Oligoadenylate synthetase see OAS
Omegasomes 117
OSBP 132, 138
Oxysterol binding proteins see OSBP

P

P bodies 35, 40
p21-activated kinase 1 see Pak1
p53 27
p62 see SQSTM1
PABP 26–27, 94, 96
Pak1 13
PAMPs 36, 104
Pan 2/3 40, 94–96, 107
Paraspeckles 93
PAS 117
Pathogen associated molecular patterns see PAMPS
Pattern recognition receptors see PRRs
P-bodies 93–108
PCBP1 35
PCBP2 28–30, 35
PERK 95
Pfeiffer, Julie 79
Phosphatase and tensin homolog see PTEN
Phosphatidylcholine (PC) 131
Phosphatidylethanolamine (PE) 115, 129–135
Phosphatidylinositide (PI) 130–131
Phosphatidylinositide phosphate lipids 131
Phosphatidylinositol 4 kinases (PI4K) 133–134, 138
Phosphatidylinositol 4-phosphate (PI4P) 129–135
Phosphatidylinositol 5-phosphate (PI5P) 132, 135
Phosphatidylserine 120

Phosphoinositide 3-kinase (PI3K) 13
Phosphoinositide 4-kinase-beta (PI4KIIIβ) 5, 133–134, 136–138
Phospholipase A2, Group XVI see PLA2G16
Phylodynamic study 83
PI (4,5) P_2 132. 135
Pichia pastoris 119
Picornavirus Study Group of ICTV 52–53, 55
PKR 95, 98
PKR-like endoplasmic reticulum kinase see PERK
PLA2G16 12, 121
Pocket factor 8, 11–12
Polio vaccine 62
Poliovirus 3, 79
  2BC 120
  2C 120
  3A ER-Golgi traffic inhibition 120
  3A membrane curvature 121
  entry 15
  G64S 72
  high fidelity variants of 72
  mutation rate of 71–72
  receptor (PVR) 9
  relative robustness 74–76
  replication organelles 130–131
  eradication 74
Poly(A) tract 24
Poly(A)-nucleases 102
Poly(C)-binding protein 1 see PCBP1
Poly(C)-binding protein 2 see PCBP2
Poly-A Binding Protein see PABP
Polypyrimidine tract-binding protein see PTB
Pre-autophagic structures see PAS
Pre-initiation complex 43S 26
Probable ATP-dependent RNA helicase DHX58 see LGP2
Processing bodies (P-bodies, PBs) 93–108
Processing bodies 25
Protein kinase B (PKB) see Akt
Protein kinase R (PKR) 94, 104–105
Protein quality control (PQC) pathways 100–101
Proteinases, viral 25
PRRs 36, 104
P-selectin glycoprotein ligand 1 see PSGL-1
PSGL-1 10–11
PTB 28–30
PTEN 135
Pyroptosis 2

Q

Quasispecies theory 3, 69–70
Qβ phage 70

R

Rab11 136–138
Rab34 14
Rab5 14
Rab7 14–16
Rab-GTPases 14–16, 136–138
Rac1 13–14

RACK1 104
Rasgap-SH3 domain binding protein 1 and 2 see G3BP1 and G3BP2
Rasgap-SH3 domain-binding protein see G3BP
Rasputin 102
RB1 inducible coiled-coil 1 see RB1CC1 and FIP200
RB1CC1 (FIP200) 118
Rck/p54 (DDX6) 103, 107
Receptor of activated protein C kinase 1 see RACK1
Receptor-interacting protein kinase 3 see RIP3
Receptors, in uncoating 7–8
Receptors, viral 9–10
Recombination 64
Recombination 73–74
Red Queen Hypothesis 71
Replication organelles 130
Replicative form (RF) 36–37
Replicative intermediate (RI) 36
Respiratory syncytial virus (RSV) 61
Retinoic acid-inducible gene-I see RIG-I
Rhinovirus 16 (HRV16) 31
Rhinovirus 1A (HRV1A) 37
Rhinovirus C 130–131
Rhinovirus replication organelles 135
Rhinovirus, subtypes 51–53
Rhinovirus, uncoating of 12
Rho GTPases 15
Ribavirin 72
Ribosomal Protein S25 see RPS25
RIG-I 36–37, 106
RIG-I-like receptor 3 (RLR-3) see LGP2
RIG-like receptors (RLRs) 36–39
Rinderpest 3
RIP3 122
RNA release 12
RNA-dependent RNA polymerases (RdRp) 24, 71, 132–133, 135
RNAse L 106
ROBETTA 57–59
RPS25 5

S

Sabin vaccine 3
Sam68 39, 97
SCARB2 10
Scavenger Receptor Class B Member 2 see SCARB2
Schuster, Peter 70
Sequence space 74–78, 83
Sequestosome1 see SQSTM1
Serine/arginine (SR)-rich protein see SRP20
Severe acute respiratory syndrome virus (SARS) 52
Sialic acid 10
Signalsome 15
Simmons, Peter 53
Sjögren syndrome 28
Small 40S ribosome 94
Smallpox 3
Smith, John Maynard 78
SQSTM1 (p62) 117, 122
Src family tyrosine kinases 15

Src-Associated substrate in Mitosis of 68 KDa see Sam68
Srp20 28–31
Stem-loop IV and V of IRES 26
Sterol metabolism 130
Stress Granule Assembly Factor 1 see G3BP1
Stress granules (SGs) 25, 35, 39, 93–108
Synonymous mutations 83

T

TAF110 26–27
TAR DNA-binding protein 43 see TDP43
TATA binding protein see TBP
TATA-binding protein-associated factor 110 see TAF110
TBP 26–27
T-cell restricted intracellular antigen 1 see TIA-1
TDP2 29–35, 41
TDP43 94, 97
TDRD3 94
TFIID 27
TFIIE 27
Theiler's murine encephalomyelitis virus (TMEV) and stress granules 102
TIA-1 39, 94–100
TIA-1-related protein see TIAR
TIAR 39
TIR-domain-containing adapter-inducing interferon-β see TRIF
TLRs 4 36, 104
TNF 2
Toll-like receptors see TLRs
Tomato bushy stunt virus (TBSV) replication organelles 130
Tomato bushy stunt virus (TBSV) replication organelles 130, 132, 135
Transcription factor II D/E see TFIID and TFIIE
Transportin import pathway 28
TRIF 106
Tristetraprolin see TTP
TTP 94–97
Tudor Domain Containing 3 see TDRD3
Tumor Necrosis Factor see TNF
Tyrosyl-DNA Phosphodiesterase 2 see TDP2

U

UBF 26
ULK1 118
Uncoating 12
Uncoordinated-like kinase 1 see ULK1
Unr protein 35
UpA dinucleotide 4
Upstream binding factor see UBF

V

Valosin-containing protein see VCP
VCP 100–101
Vesicular stomatitis virus (VSV), Population dynamics of 77, 79
Vesicular-tubular organelles 130
Vimentin 10
Virus entry 7–17
VLA-2 protein 16
VP0 capsid protein, cleavage 120